Sustainable Development for Engineers

A HANDBOOK AND RESOURCE GUIDE

Edited by Karel Mulder

SUSTAINABLE DEVELOPMENT FOR ENGINEERS

A HANDBOOK AND RESOURCE GUIDE

EDITED BY
KAREL MULDER
DELFT UNIVERSITY OF TECHNOLOGY,
THE NETHERLANDS

Greenleaf
PUBLISHING

2006

© 2006 Greenleaf Publishing Ltd

Published by Greenleaf Publishing Limited
Aizlewood's Mill
Nursery Street
Sheffield S3 8GG
UK
www.greenleaf-publishing.com

The paper used for this book is a natural, recyclable product made from wood grown
in sustainable forests; the manufacturing processes conform to the environmental
regulations of the country of origin.
Printed in Great Britain by William Clowes Ltd, Beccles, Suffolk.
Cover by LaliAbril.com.

British Library Cataloguing in Publication Data:
A catalogue record for this book is available from the British Library
ISBN-10: 1-874719-19-5
ISBN-13: 978-1-874719-19-9

Contents

Acknowledgements

This book has been written as a joint project of Delft University of Technology (DUT) and the Technical University of Catalonia (UPC) in Barcelona. Karel Mulder has been the primary author. He is indebted to many colleagues that submitted material, gave advice and commented on drafts:

Carlos Welsh, Josep Antequera, Bárbara Sureda and Ernesto González from the Sustainability Observatory of the UNESCO Chair on Technology, Sustainable Development, Imbalance and Global Change, UPC, submitted the first draft of Chapter 6.

Renate van Drimmelen (DUT) submitted a first draft of Chapter 7 and Heleen Willems (DUT) submitted a first draft of Chapter 8.

Carlos Welsh, Enric Carrera, Didac Ferrer Balas, Yazmin Cruz, Mireia de Mingo, Marta Pujadas, Jordi Segalas, Albert Cuchi, Ana Luz Saco (UPC), Kai Hillebrecht (TU Braunschweig), Francisco Lozano (Tecnológico de Monterrey), Fernando Guttierez (Universidad Politécnica de Madrid), Paul Weaver (University of Durham), Leo Jansen, Linda Kamp, Gert Jan de Werk, Renate van Drimmelen, Jaco Quist, Mariette Overschie, Dirk-Jan Peet, Crelis Rammelt, Karin van Duijn, Agnes Dokkuma (DUT), and about 100 participants in a special workshop at EESD II, Barcelona, 2004, contributed to the work on this book.

Introduction

Engineers are future builders. They shape the world through their product and process designs, their management of technical systems and their innovations. It is the task of the engineer— in co-operation with other disciplines—to meet the needs of our society.

The way in which our industrial society operates is no longer sustainable and also threatens opportunities for the underprivileged and for future generations to develop. Industrial society is responsible for the depletion and poisoning of our planet Earth and has caused us to take risks of unknown magnitude. Natural systems have been altered beyond the point of no return. The unequal distribution of the Earth's resources creates unbearable tensions between the 20% rich and the 80% poor of our planet.

Society is on a changing course. It will have to adapt, and engineers, the builders of the future, can play an essential role in making it possible.

Change is needed in every area. The way we consume cannot continue. However, it is not easy to relinquish the miracles of the consumer society. Is there a way to reconcile our modern lifestyles with the compelling need for change? Could new, improved technology play a key role? Great leaps in the environmental efficiency of our current technologies would be needed. Can they be realised? And are engineers trained to undertake this important task?

In most industrialised countries, the engineer is envisaged as a bright individual inventing or designing new products, processes or public works at his[1] drawing board. However, his inventiveness might just be channelled too much towards developing clever technology, and too little towards societal needs. Scientifically trained engineers must focus on questions of sustainability and how social and environmental issues impact on technology. After graduation, engineering students do not live in ivory towers, but become managers of design or research groups, leaders of companies or institutes and perhaps also future policy-makers. Engineers should therefore learn to think strategically about the sustainability challenges of technological change. Of course, that does not imply that each individual engineer is constantly engaged only with sustainable development. However, he

---

[1] The number of female engineers varies, but in almost every engineering discipline, and in most countries, female representation is far below 50%. This is deeply regrettable. For simplicity we refer to the engineer here as 'he', although it is to be hoped the situation changes soon.

should at least be aware of the challenges it poses to his profession and be able to communicate with other disciplines in developing solutions.

This book has been written to give the engineering student insight into the challenge that sustainable development poses to the engineering profession, the contribution of engineering to sustainable development, and the barriers and pitfalls to beware. The engagement of engineers in sustainable development is good for sustainable development, and good for the engineer who wants to broaden his perspective.

1 Why do we need sustainability?

This chapter explains what sustainable development is and the consequences of unsustainable practices. It does this by sketching the history of Easter Island—an example of the decline of a community caused by its unsustainable practices. The chapter describes briefly some of the history of the concept of sustainable development and offers a framework for setting targets for future improvements.

Easter Island

Easter Island (Rapa Nui) in the Pacific Ocean is one of the most remote inhabited places in the world. It is 3,747 km from Chile (to which it belongs) and 2,250 km from the nearest inhabited island, Pitcairn. Its enormous statues (see Figure 1.1) make it a popular tourist attraction.

The Netherlands admiral Jacob Roggeveen landed on the Easter Island in 1722. On his journey around the world, he was in search of a large mysterious island that was said to be in the Pacific.

When Roggeveen and his crew explored the island they found a population of about 3,000 people living in shacks and dressed in rags. Roggeveen saw the large statues (called Moai) and was intrigued by them. The statues were placed in groups on platforms (Ahu) near the shoreline and all faced inland. Some had separate stone hats. The largest standing statue weighs 82,000 kg (82 tonnes) and is 9.8 m. The statues were cut in a quarry in the middle of the island up to 14 km from the shoreline. A number of semi-finished statues can still be observed at this site together with several finished statues that did not make it to the shoreline.

FIGURE 1.1 Moai at Easter Island

Roggeveen stayed for only a couple of days on the island. Later explorers described the island life in more detail and puzzled over the statues. Creating the platforms, cutting the statues and transporting them over several kilometres must have taken considerable effort. There are means of transporting an average statue using about 20 people; transport would then take 30–70 days. However, these methods are risky on the rough terrain of Easter Island. More secure methods of transport would use rollers or sledges. These methods are faster but require more people.[1]

Two things are clear:

● Creating the statues required a great deal of labour. As most statues date from between 1400 and 1600, Easter Island society would have

1 For modern theories, see www.anthropology.hawaii.edu/projects/ppp/symposium. html, 20 October 2005. For transport experiments on Easter Island, see www. pbs.org/wgbh/nova/easter/move, 20 October 2005.

had to be far more prosperous and far better organised at that time to be able to support these efforts

- Transportation methods that did not use wood would have been very risky for the statues and probably inadequate on the rougher terrain

Another piece of evidence points to the use of wood: the island was covered by bushes and not by trees. Only two Chilean wine palm trees have ever been discovered on the island; they were growing in a canyon and could not be reached. However, archaeologists have proved that the island was once covered by various species of palm trees.

Modern theories regarding Easter Island state that its inhabitants came from Polynesia and arrived there between 400 and 800 AD.[2] The richness of the land and sea gave the islanders the means to develop a rich culture. The population grew to a level of approximately 7,000 in the 16th century. As the Polynesian social system is based on the clan as dominant unit, clan life was probably the dominant social system on Easter Island too. The Ahu were probably religious symbols that also expressed the status of its owning clan. But what happened?

The growing population and the enormous activity deployed in creating Moai took their toll. The fertile soil eroded and more land was needed for agriculture. Large numbers of trees were cut for building Moai, boats and houses. The island probably ran out of wood and food, but the islanders must have continued chopping trees until the last available one. Consequently they were unable to continue the creation and placement of Moai; they could not build boats anymore and construction of new houses became impossible. The accepted hypothesis is that this coincided with armed conflict between the clans.

The civilisation of Easter Island collapsed because of its unsustainable nature. But why didn't anyone do anything about it? Maybe the inhabitants were not as aware, as we are now, about the harm they were inflicting on the ecosystem of their island. It is unrealistic to assume that a civilisation as highly developed as that on Easter Island did not appreciate the fact that their last tree had been cut down. But in the competitive struggle among clans, not cutting down a tree would mean leaving it for the axes of a competing clan. This is an example of a **prisoner's dilemma** (see Chapter 4).

---

2 Recent evidence suggests that the colonisation of Easter Island occurred much later, i.e. in the year 1200. It is also suggested that the devastation of the island happened much more quickly. See Ann Gibbons, 'Dates Revise Easter Island History', *Science* 311.5766 (10 March 2006): 1,360.

Unsustainable societies collapse

The example of Easter Island shows that people cannot act as if their resources are limitless. But Easter Island is not an extreme example. Various civilisations throughout the world have collapsed as a result of their use of unsustainable forms of agriculture. To cite only a few examples:

- When irrigation is used, agriculture yields good harvests in the beginning. But when the water evaporates on the land, it leaves salt in the top layer. Plants are poisoned by too much salt; soon yields start to decline and the land can no longer be used for agriculture. There is evidence that this phenomenon contributed to the collapse of the civilisations of the Maya in Central America, the Indus Valley in southern Asia and Mesopotamian cities in the Middle East

- When agricultural land is drained, hydrostatic pressure can force saline water to rise from lower layers. This can also lead to salt poisoning

- Irrigation often contributes to erosion, removing the productive top layer of soil from the agricultural land. Erosion can also lead to dust storms, which can themselves pose a threat to farms and villages. In the 1930s, the so-called Dust Bowl, caused by years of low rainfall and extensive agricultural production, created a severe problem in the Midwest of the USA. Throughout the world, communities have been destroyed by land erosion

- When the population grows, the increasing need for food can lead to extensive monocultural (single-crop) land use. This can lead to epidemics of plant disease (especially without any crop rotation or fallow periods). The resulting famine may destroy civilisations. The famine that caused Abraham's descendants to leave the land of Canaan for Egypt was probably due to wheat rust.[3] A well-known example is the Irish Potato Famine between 1845 and 1850.[4] Ireland was a densely populated nation in 1845. To produce enough food, it depended almost completely on its potato crop, as potatoes have very high yields per area. Crop rotation to prevent explosions of soil-related plant diseases was rarely possible. The blight (*Phytophthora infestans*) first struck in 1845. It left the potatoes rotting in the fields. Stores were also affected. The blight struck again in 1846 and 1848. About 1 million people died in the famine and many fled the country, causing a drop in the Irish population from 8 million to 5 million

— Just Mother Nature re doing humeranes the Census! (Aids / Plague / Malaria) ect.

3 www.botany.hawaii.edu/faculty/wong/BOT135/Lect08.htm, 20 October 2005.
4 collections.ic.gc.ca/potato/history/ireland.asp, 20 October 2005.

But how about us?

Nowadays world oil consumption is about 30 billion barrels annually. Proven reserves amount to some 1,150 billion barrels; this is sufficient for 25–35 years depending on the increase in energy consumption.[5]

New reserves are explored especially if oil prices rise. However, there will be a point when no new reserves are available. If we continue consuming oil at the current pace, it will probably run out somewhere between 2050 and 2100. Oil is crucial for our transport systems, electricity supply, heating and materials supply. We may recognise the problem but will we be able to take action in order to develop the alternatives we need? Technologies that use tar sands or coal might provide alternative fuels but what about climate change? Scientists are gradually reaching consensus on this phenomenon, but can the global community reach consensus on actions to take? The aftermath of the Kyoto conference on climate change shows that consensus on worldwide measures is hard to reach (cf. Chapter 4).

The Easter Island story contains another lesson: using the dichotomy between technology and society is not very productive. The Easter Island technology was invented to produce statues that were important in its culture and social organisation. Changing the technology would have had serious consequences for the island's culture and organisation. A change in technology always needs to be attuned to social processes, while a change in social structures will have consequences for technology. We therefore need to address **socio-technical change**.

The crisis that we face today has both environmental and social components. Many remain unconcerned about this crisis. On the one hand, they have an insufficient perception of its magnitude—especially in the developed countries. On the other hand, the consequences are hard to accept.

> We have acquired the capability to disturb the Earth's natural systems, but we do not want to accept the responsibilities of this practice.[6]

> Perhaps more dangerous for the environment's integrity is our way of perceiving the threats than the threats themselves. Most people resist accepting the crisis' extreme seriousness.[7]

Nowadays we have access to a great deal of information about the world situation. If humanity fails to act, it will not be because of a lack of information.

5 For precise data, see BP *Statistical Review of World Energy 2005*; www.bp.com/genericarticle.do?categoryId=111&contentId=2004175, 20 October 2005.
6 L.C. Brown (ed.), *State of the World 1999* (Washington, DC: Worldwatch Institute, 1999; www.worldwatch.org/pubs/sow/1999, 20 October 2005).
7 A. Gore, *Earth in the Balance: Ecology and Human Spirit* (New York: Plume Books, 1993).

W.T.F. has heard of ✓

- The UN Environment Programme (UNEP) regularly publishes data sets that describe the state of the world[8]

- The Worldwatch Institute publishes an annual *State of the World* report and a bimonthly *Worldwatch Magazine*[9] *Everyone gets one*

- There is widespread media coverage of international conferences *(ipH!)* such as:
  - The United Nations Conference on Environment and Development in Rio de Janeiro in 1992[10] and in Johannesburg in 2002[11]
  - The International Conference on Population and Development in Cairo in 1994 and in New York in 1999[12]
  - The Fourth World Conference on Women in Beijing in 1995[13]
  - The World Food Summit in Rome in 1996[14]
  - The World Summit on the Information Society in Geneva in 2003[15]
  - The Conference on Human Settlements Habitat in Istanbul in 1996[16]

½ population is young or doesn + 1/20 of them most likely don't watch the news for info. ✗

Sustainable development?

The natural environment is the source of all substances that sustain human life. We take food, water, fuels, minerals and metals from it and we use it as a receptor ('sink') of our waste.[17] The general attitude towards the environment has changed sharply over recent decades. In the 1960s, most people perceived the natural environment as infinite. In 1962, Rachel Carson showed in her book *Silent Spring*[18] that the use of agricultural chemicals had created severe problems. In particular one modern pesticide, DDT, which

8 United Nations Environment Programme (UNEP) Global Environmental Outlook, www.unep.org/Geo/index.htm, 20 October 2005.
9 www.worldwatch.org/pubs, 20 October 2005.
10 www.ciesin.org/datasets/unced/unced.html, 20 October 2005.
11 www.johannesburgsummit.org/html/basic_info/unced.html, 20 October 2005.
12 www.iisd.ca/cairo.html, 20 October 2005.
13 www.un.org/womenwatch/daw/beijing, 20 October 2005.
14 www.fao.org/docrep/003/w3613e/w3613e00.htm, 20 October 2005.
15 www.itu.int/wsis, 20 October 2005.
16 www.un.org/Conferences/habitat, 20 October 2005.
17 What counts as 'waste' may change by place and over time. Manure might be a fertiliser in small quantities, but a tremendous environmental problem in larger quantities. In richer countries, plastic packaging is generally a waste after it has been used. In many poorer countries, it is a resource of value.
18 R. Carson, *Silent Spring* (Boston, MA: Houghton Mifflin; Cambridge, MA: Riverside Press, 1962).

✗ We are in a extremly developed Cantry + we are so comfy eh ✗ worry about this - we have

was widely used to kill insects especially because of its low toxicity to humans and livestock, turned out to have disastrous effects on wildlife. The cause was found to be the accumulation of DDT in the fatty tissue of predator animals.

Various incidents contributed to the public's growing awareness of the environment. They included:

- The soil contamination of the Love Canal site in Niagara Falls (USA) where a school had been built on a chemical dump site[19]

- The enormous oil spills following the first ever running-aground of a supertanker, the Torrey Canyon, near the Scilly Islands in the UK in 1967[20]

In 1972, Dennis Meadows and his fellow authors took a wider perspective in Limits to Growth.[21] The book marked a change towards a finite vision of the world; it focused not just on waste and emissions but also on resource consumption. The book observed that both population and consumption growth were exponential while resource production can grow only in a linear fashion. Factors such as population growth, resource consumption, food production and pollution were integrated in a single model, which predicted a collapse before the year 2000. The pessimistic tone of the book triggered a reaction from American futurist Herman Kahn, who envisaged a brighter future:

> Two hundred years ago almost everywhere human beings were comparatively few, poor and at the mercy of the forces of nature, and 200 years from now, we expect, almost everywhere they will be numerous, rich and in control of the forces of nature.[22]

The basis for this vision was technological improvement. Kahn predicted that humankind would exit from the track of exponential growth to end up at a final level in which every world citizen could live a prosperous life. Given the environmental crises described above, this bright future is by no means certain. It is a challenge to which we need to direct our efforts.

In the 1980s, the UN World Commission on Environment and Development presided over by Norway's prime minister Gro Harlem Brundtland introduced the term 'sustainable development' to designate the challenges for the future development of our planet:

19 See ublib.buffalo.edu/libraries/projects/lovecanal, 17 March 2006.
20 See www.imo.org/Environment/mainframe.asp?topic_id=231#2, 17 March 2006.
21 D.H. Meadows, D.L. Meadows, J. Randers and W.W. Behrens, The Limits to Growth (New York: Universe Books, 1972).
22 H. Kahn, W. Brown and L. Martel, The Next 200 Years: A Scenario for America and the World (New York: William Morrow, 1976).

toxic waste dumps, pollution!! even population!

> Sustainable development is a development that meets the needs
> of the present without compromising the ability of future gen-
> erations to meet their own needs.[23]

In the report, the Commission proposed reconciling the development issue with the protection of the planet's resources.

Sustainable development is about reaching new equilibria:

- Between the poor and the rich *U need Xing Kang*
- Between current and future generations *) goes for Kang*
- Between humankind and nature *most important*

Sustainable development is a direction in which to proceed; nobody can claim the wisdom of defining an ultimate solution for the world.

Sustainability does not imply that everybody will end up living in exactly the same way as everybody else on this planet. Needs differ between cultures and between individuals. Cultural diversity is worth preserving because it is our living heritage.

greed But why should the richer part of the world's population be able to fulfil *egoism* even the most absurd desires while many others are starving, unable to attend school or dying of diseases that are easily curable in richer parts of the world? Sustainable development is a moral issue built on the assumption *ha* that all human beings are born with equal rights to build the life they choose *ha* (without harming the rights of others). Human rights, gender issues, employment and cultural traditions cannot be separated from the sustainability issue. Countries, regions and local communities should develop their own path towards sustainability; there is not just one route. Any attempt to develop sustainable directions of development for others should therefore be judged with suspicion (see Chapter 3).

What is the basis of sustainability?

Although we are not able to define a sustainable society as a final situation to strive for, some more basic principles can be described:

- Resource consumption should be minimised
- Cycles of consumption of non-renewable materials should be closed
- Preference should be given to renewable materials and energy sources

23 G.H. Brundtland (ed.), *Our Common Future: The World Commission on Environment and Development* (Oxford: Oxford University Press, 1987).

- Development of human potentials such as communication, creativity, co-operation, intellectual development and love should be stimulated

- One should contribute to the common good and not just to the private good

Unsustainable activities can be defined as those that:

- Require a constant consumption of non-renewable resources or consume more renewable resources than the Earth system might generate

- Cause degradation of the environment

- Require such quantities of resources that they will never be available for everyone

- Bring species to extinction

- Stimulate selfishness

- Create the risk of a disaster

Sustainable development implies a redefinition and review of concepts such as production, wealth and interest. Economic theory should find ways to include the assets of nature and human development into its calculations. National and international law, taxation and commercial practice need considerable change.

Can we afford to switch to sustainable development? There is no alternative for sustainable development. The world has to proceed in that direction or it will face chaos and decline. Moreover, the developed world will have to make a start and should be aware that its affluence is due to an **ecological debt**.

Sustainable development will create more employment. For example, energy conservation and organic farming are far more labour-intensive than industrialised agriculture and either fossil fuel or nuclear power production.[24] However, individual countries or regions could suffer considerably if they opted completely for sustainable development. For example, sustainable fisheries might imply fishing much less to maintain a rich and productive ecosystem (see Chapter 2). But foreign fishermen might take over fishing grounds and fish markets, thus preventing sustainable fisheries from taking effect and putting sustainable fishermen out of work. Such a 'sustainable development' course would be a dead-end street.

For sustainable development, we need co-operation and international agreements. Agreements that take into account the rights and needs of all.

24   B. Commoner, *Making Peace with the Planet* (New York: Alfred A. Knopf, 1960).

The role of technology: factor X

Technology will play an important role in sustainable development. However, the sustainable technology is not going to be the one that chops the trees faster and more efficiently, but the one that is able to increase the useful lifetime of the trunks.

What improvements in the environmental efficiency of technologies do we need?

In the 1970s, there were debates about which factors contributed most to the problems we were facing— growth in consumption, overpopulation or the state of technology. A rough relationship between these factors can be described by the so-called IPAT equation:

$$I = P \times A \times T \qquad\qquad (1)^{25}$$

where:
I = Total environmental impact of humankind on the planet
P = Population
A = Affluence, number of products or services consumed per person (i.e. for economists the annual Gross National Product per capita)
T = Environmental impact per unit of product/service consumed

T is often called the factor 'Technology efficiency'. But it is important to note that T diminishes as technologies become more efficient! Moreover, T also reflects non-technological issues such as product re-use and the organisation of production.

The IPAT equation can be used to obtain clarity on the magnitude of the environmental technological efficiency improvements that we need to reach in the long term. We therefore need estimates of the various factors.

- **Environmental impact**. As explained in Chapter 2, our current use of natural resources is unsustainable. Assume that we want to cut it by 50%

- **Population growth** has been exponential. In 2000, world population was approximately 6 billion. In the past decade, we have seen declining population growth rates. This is especially due to the devastating effects of the HIV epidemic. In large parts of Africa and on an increasing scale in Asia, up to half the teenagers are HIV-positive. Not only the direct death toll is important, but also the fact that youngsters do not reach the age of reproduction. Population growth is hardly affected by government policies. Even large-scale wars barely influence it. Only long-term policies can stabilise the global population. Growing affluence contributes considerably to a

---

25 P. Ehrlich and J. Holdren, 'Impact of Population Growth: Complacency concerning this component of man's predicament is unjustified and counterproductive', *Science* 171 (1971): 1,211-17.

Area	2000	2050
World	6,071	8,919
More developed regions	1,194	1,220
Less developed regions	4,877	7,699
Least developed countries	668	1,675
Africa	796	1,803
Asia	3,680	5,222
Europe	728	632
Latin America and the Caribbean	520	768
North America	316	448
Oceania	31	46

TABLE 1.1 **World population forecast (millions)**

Source: UN Department of Economic and Social Affairs, Population Division, *World Population in 2300*, March 2004, www.un.org/esa/population/publications/longrange2/longrange2.htm, 25 October 2005.

decline in population growth. The global population in the year 2050 is predicted to be between 8 and 11 billion (see Table 1.1). Therefore, a rough estimate of population growth is a factor of 1.5[26]

- **Affluence.** The richest 20% of the world population is consuming roughly 80% of the world's resources. This leaves only 20% for the remaining 80% of the world population.[27] The rich people in the world therefore consume, on average, 16 times more resources. The economies of the developed world are growing on average by 2% annually. Over a 50-year period, this implies a growth factor of 2.7. If the poorer nations want to catch up with the richer nations, they need to grow by a factor 16 × 2.7 = 43.2, which means an annual growth of 7.8%. The combined growth of the poorer and richer parts of the population can then be calculated. Let us assume consumption now is 100. The rich consume 80 and the growth factor is 2.7 so consumption in 50 years' time will be 216. The poor now consume 20 and the growth rate is 43.2, so consumption in 50 years' time will

26   F. Pearce, 'Global Population Forecast Falls', *New Scientist*, 27 February 2003.
27   For more precise figures, see United Nations Development Programme (UNDP), *Human Development Report 2003. Millennium Development Goals: A Compact among Nations to End Human Poverty* (New York: UNDP, 2003, hdr.undp.org/reports/global/2003, 20 October 2005).

be 864. This gives a total consumption of 1,080 or 10.8 times the starting level.

We can substitute these estimates into the IPAT equation. For the year 2000, we can set all factors at a reference value of 1:

$$I_{2000} = P_{2000} \times A_{2000} \times T_{2000} = 1 \times 1 \times 1 = 1 \tag{2}$$

Now we can calculate the value of T_{2050} ($X \times T_{2000}$) by substituting the estimates made above:

$$I_{2050} = P_{2050} \times A_{2050} \times T_{2050} = 1.5 \times P_{2000} \times 10.8 \times A_{2000} \times X \times T_{2000} = 0.5 \times I_{2000} \tag{3}$$

We can now calculate X:

$$X = 0.5/(1.5 \times 10.8) = 1/32.4 \tag{4}$$

Based on the above assumptions, technology should thus be 32.4 times more environmentally efficient than it is today.

However, these calculations should not be interpreted literally as a target for each separate technology.

As explained in Chapter 5, technological change always leads to social changes. Needs are dynamic and influence technology; new technologies change needs.

Secondly, the world will never be 'finished'. Sustainable development must be an open process in which future generations (and underprivileged people that have no voice now) can join in and set new targets. Thus, there will always be new problems and new challenges, even if we have improved the environmental efficiency of technology by a factor of 32.4.

Thirdly, some technologies can be improved only marginally while other technologies can be improved radically to yield much more than a 32-fold improvement without any consumption of resources or emissions of waste. Marginal and radical technological improvements will both be important for sustainable development. But we have to stimulate the radical ones because, only when we produce sufficiently large leaps in technological efficiencies can we reach the orders of improvement needed. This is explained further in Chapters 9 and 10.

Questions, discussion and exercises

1. The IPAT equation describes the environmental impact as a linear function of the factors Population, Affluence and Technology.
    a. Do you think this description is realistic? Why do you think so?
    b. Consider the case in which the Impact depends on the square of the Population. How does this affect factor X?

    c. Suggest which of the factors (Population, Affluence and Technology) are interrelated (i.e. they are a function of one of the other factors). Give reasons why you believe the interrelation exists.

2. Calculate the required improvement in the environmental efficiency of technology in 2050 assuming that the affluence of the whole world population is at the level of affluence of the rich countries in 2000.

3. Shortages in the supply of oil are claimed to be a major threat to the world's future. Describe another threat that could potentially lead to catastrophe and analyse its consequences.

4. Search the internet for other major collapses of cultures than those mentioned in the text. What were the reasons for the collapse? Was the unsustainable behaviour the cause of the problem? How was this behaviour embedded in the culture?

5. Some environmental disasters received media coverage only in the nations in which they occurred. Try to find out which environmental disaster was especially important in creating environmental awareness in your country.

2 Why is the current world system unsustainable?

This chapter examines the consequences of current world consumption and production systems for the various natural systems that sustain life on Earth. An analysis of the natural system that sustains life on this planet shows that it is constantly evolving. Humankind is part of this system. Our perception of the universe, which is often rooted in religious or political convictions, is important to determine what we can do to the planet. The chapter includes a discussion of criteria that can be used to determine which of the changes inflicted on the life-sustaining system by humankind should be regarded as problematic and an overview of the main problems.

The life-sustaining system

An **ecosystem** is the basic functional unit that sustains life. An ecosystem includes the plant and animal communities in an area together with the non-living physical environment that supports them.[1] A major problem is identifying the borders of an ecosystem: there are always interchanges between two adjacent ecosystems.

---

1   US Environmental Protection Agency; www.epa.gov/maia/html/glossary.html, 21 October 2005.

Ecology is the science that deals with ecosystems. Ecology is an interdisciplinary science with connections to various disciplines such as meteorology, climatology, hydrology, biology, genetics, physiology, aetiology and mathematics.

Ecology studies the evolution of life at three integration levels:

- Dynamics of the population of a single species

- Dynamics of communities of species

- Dynamics of ecosystems (community + biotope)

The 'balance' paradigm dominated ecology for a long time. This paradigm supposed that ecological systems were:

- Closed

- Free from disruptions and alterations

- Independent of human influence

- Regulated by their own mechanisms, which help to achieve stability and balance

This paradigm has now been renounced following increasing evidence that ecosystems are basically open systems or that balance exists only at a much higher systemic level.

Interactions between species

In an ecosystem, the different species and organisms are not isolated. Organisms are related to one another with various positive or negative interactions as follows:

- **Mutualism or symbiosis**: when species obtain benefits from their association. Examples include birds that spread the seeds of plants they eat, insects that are vital for flower reproduction, and birds that remove parasites from the skin of mammals

- **Commensalism:** when one of the species receives a benefit and the other species is not affected. Scavengers are a good example

- **Competition:** species may compete, for example, for food, such as two species that eat the same plant. If two species aim to occupy the same place in the energy or materials exchange of an ecosystem, there is competition

- **Depredation:** when a species (predator) eats another species (prey). The sizes of the predator and prey populations are related, as the success of the predator creates a shortage of food that leads to starvation

- **Parasitism** is a special case of depredation. In this case, the predator tends to keep the prey alive as it feeds upon the living prey

To safeguard their presence in ecosystems, individual species apply various strategies:

- **Reproductive strategy.** This ensures the species survives through a high rate of reproduction. Many individual members of the species die prematurely (before reproduction). This strategy is used in harsh environments with variable weather conditions and frequent disruptions. The strategy can be observed in early colonists (species that conquer a new territory) and in opportunist species (e.g. rats, mosquitoes and rabbits)

- **Individual survival strategy.** This aims for a high survival rate (but low reproduction). This strategy can be observed in stable and almost constant environments. The individuals are, to a high level, independent of their environment. Examples include elephants, whales and, of course, humans

Ecosystems as energy-processing units

A more or less constant flow of energy is indispensable to maintain the vital functions of the living species. An ecosystem can also be considered an energy-processing unit. The energy enters the ecosystems through primary producers or autotrophs, i.e. plants or organisms that are able to produce organic substances from simple inorganic substances such as carbon dioxide (CO_2) and water with the help of sunlight (ultraviolet radiation).

To analyse the transport of energy in an ecosystem, we can divide it into trophic levels (Table 2.1).

This distinction between trophic levels is a simplification of the ecosystem. In fact, most of the organisms are extraordinarily adaptable, changing their feeding habits depending on the circumstances or their own life-cycle.

The different trophic levels are interdependent. This constitutes a 'food chain'. In every ecosystem, there are interconnections between different food chains, e.g. the same plant can be eaten by many different consumers or an animal can eat many different plants or animals. Thus, an ecosystem is not just a bundle of food chains but consists of trophic or food webs (see the example in Figure 2.1).

Heterotrophs oxidise organic material through a process called aerobic combustion (carbohydrates react with oxygen to produce CO_2 and water). The energy is stored in organic molecules and used by the organism for different functions (e.g. movement, reproduction, etc.). The consumption of energy ultimately involves changing it from chemical energy to thermal energy, which cannot be used by the biosphere.

Disposable energy levels fall in every level of the food chain due to thermal energy losses.

Tropic level	Description
Primary producers or autotrophs	These organisms can be divided into: ● Phototrophs: primary producers that use sunlight as an energy source. They store it by making complex molecules. Plants, algae and photosynthetic bacteria are all phototrophs ● Chemotrophs: primary producers that use chemical reactions to obtain energy. Chemical sources might be hydrogen sulphide or sugars
Consumers or heterotrophs	These organisms are unable to produce their own food. They feed on: ● Primary producers (herbivores) ● Other consumers (carnivores) They usually feed themselves by oxidising organic material
Decomposers	These organisms feed on the remains of diseased consumers. Various groups can be discerned, e.g. scavengers and microorganisms

TABLE 2.1 Division of an ecosystem into trophic levels

FIGURE 2.1 Example ecosystem

Source: regentsprep.org/Regents/biology/units/organization/ecosystem.gif; © 1999–2003 Oswego City School District Regents Exam Prep Center

The ecosystem is not an isolated entity; it receives inputs and outputs of substances and energy. The biological system consumes endosomatic energy (internal energy from the organism's metabolism), which comes in from the primary producers and moves through the food webs.

Humans also consume exosomatic energy (energy that does not pass through our bodies). Nowadays, we basically use exosomatic energy (e.g. oil and gas).

Since the Palaeolithic period (Old Stone Age), endosomatic energy consumption has increased threefold but exosomatic consumption has increased by a factor of 110.

Material flows in ecosystems

Apart from energy, organisms need substances with which to create cells and tissue. Plants and living organisms are mostly formed by water and organic molecules such as lipids, glucose and proteins.

Substances are used in cycles (basically due to the decomposers in the ecosystem). Closed cycles of carbon, oxygen and nitrogen are essential to support life on Earth.

Ecosystems can be characterised by the quantity of biomass that they contain and the energy that is consumed in the system.

- **Biomass** is the total mass of all the living organisms in an ecosystem. Biomass can be measured for the total ecosystem, but also as a relative density (i.e. grams of carbon [C]/surface or volume unit). In land ecosystems, the most usual unit is gC/cm^2 or gC/ha and, in aquatic ecosystems, it is gC/cm^3

- **Production** is the amount of new biomass that is produced by the primary producers per time unit. It represents the energy that is disposable for the next trophic level and can be measured in $gC/cm^2/year$

If the amount of energy that a superior trophic level takes from a lower level is larger than the production of the lower level, then the production of the lower level will be affected. This could lead to destruction of the ecosystem. The production of an ecosystem is therefore a basic parameter to define its capacity.

- **Productivity** is the quotient of production and biomass (i.e. production divided by biomass). It gives an approximation of the relative speed of production of the ecosystem

- **Renewal time** is the quotient of biomass and production (i.e. biomass divided by production) and indicates the time needed to renew the entire biomass

Carbon, sulphur and nitrogen cycles

Because organisms are composed of water and carbon (organic) compounds, the carbon cycle is vital to maintain life on Earth. Carbon is found in the:

- **Atmosphere**: contains little carbon (350 mg/l), but forms the main intermediate between carbon in other systems

- **Oceans**: contain much carbon in the form of inorganic soluble carbon, organic soluble carbon and organic carbon (living creatures, plants and their remains). The main carbon storage of oceans is in sedimentary rocks

- **Biosphere and lithosphere**: the amount of carbon and its distribution are not clear

The basic mechanisms that close the carbon cycle are photosynthesis and aerobic combustion. In photosynthesis, atmospheric CO_2 reacts with water to synthesise carbohydrates using sunlight as the source of energy. In aerobic combustion, the carbohydrates are turned into water and CO_2—a process that releases energy to the organism.

Sulphur is a secondary component for organisms and can be found in some of the amino acids making up proteins. The largest storage of sulphur is in the Earth. Even though sulphur dioxide (SO_2) is a severe gaseous pollutant, there is only a very small quantity of sulphur in the atmosphere. Sulphur is washed out of the atmosphere by rain.

Atmospheric sulphur originates from natural emissions (microbiotic processes and volcanoes) and industrial activities. The sulphur cycle has been increasing significantly in volume since the beginning of the industrial revolution.

Nitrogen is an important nutrient for ecosystems and constitutes more than 78% of the Earth's atmosphere. Almost all plants, animals and microorganisms depend on nitrogen compounds such as ammonium, nitrates and nitrites, which are rare in soil and water.

Nitrogen enters the cycle by biological fixation. Nitrogen compounds of organic origin are released into the environment by fungi and bacteria. In the ammonification process, organic nitrogen is transformed into inorganic compounds or minerals. Ammonia can become nitrite and, afterwards, nitrate, through the nitrification process.

Bacteria, plants and fungi can use nitrates and ammonium compounds directly in the assimilation process. Finally, nitrogen is released from the organic material of an ecosystem through denitrification.

Ecosystems and change

Most organisms are restricted physiologically by specific environmental conditions such as temperature or the availability of water. These conditions determine the distribution and abundance of species. Diversity in an ecosystem helps to stabilise it during adverse conditions. This ensures the global survival of a system but does not guarantee the survival of individual species.

The evolution of ecosystems is path-dependent, i.e. the wealth and diversity of an ecosystem depend on the historic process by which it was formed. Today's situation is also determined by factors such as climate and soil type. The wealth and diversity of an ecosystem may change continually.

Ecosystems can colonise new areas and become extinct elsewhere, with historical records showing a sequence of ecosystems in specific areas. Fjords and glaciers are places where this sequential development can be most easily observed. For example, the Tracy Fjord glacier in Alaska has withdrawn 40 km since 1960. By doing so, it created scope for new ecosystems. In this environment, the initial colonisation was made by lichens. Small leaf vegetation then moved in and finally a complete ecosystem formed. This can be observed close to the shoreline where a forest has begun to form.

At the same time as the number of species increases in such situations, the complexity of the biological interactions between different species also increases. More mature ecosystems generally have a more complex organisation.

The population of more diverse ecosystems tends to be more stable. Mature ecosystems such as the tropical rainforests are highly stable and are able to return to their equilibrium even under severe disturbance.

Problems in the life-sustaining system

What do we actually mean when we speak about sustainability problems or environmental problems? At first glance, the question might look simple. But, if we try to answer it from the ecosystems perspective outlined above, it is not easy to give a straightforward answer. Humankind as part of a complex ecosystem is evolving and revolutionising this ecosystem. Evolution of ecosystems happens continually. So why care? Not every disturbance of an ecosystem is a problem, but how do we recognise when it is?

Environmental problems and natural catastrophes

Problems in the life-sustaining system, which we call environmental problems, are caused by humankind. Natural catastrophes may endanger the life-sustaining system but they are not considered environmental problems though their contribution to problematic situations can be enormous. For example, the enormous amounts of chloride that were emitted during the eruption of the Mount Pinatubo volcano in the Philippines in June 1991 contributed significantly to the thinning of the ozone layer. Moreover, the volcanic aerosols reflected sunlight which created global cooling for about two years.

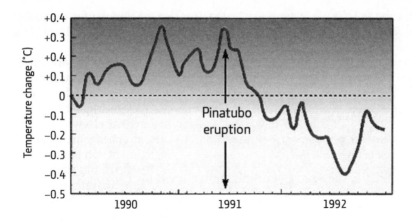

Figure 2.2 **Temperature changes at Mount Pinatubo**

Source: Steven Earle, Geology Department, Malaspina University College, Nanaimo, British Columbia, Canada; www.mala.bc.ca/~earles/pinatubo-photosynthesis-mar03.htm, 25 October 2005

The same holds for phenomena created by the ordinary dynamics of ecosystems. For example, it was recently discovered that salt marshes in warmer regions produce a considerable amount of the ozone-depleting substances methyl bromide and methyl chloride.[2]

Environmental problems are created by human activity. Calamities are supposed to be part of nature. Although the difference might seem pedantic, its importance can be inferred from our attitude: there is an implicit understanding that natural calamities are less likely to irreversibly change the life-sustaining system than are human-induced problems. Natural calamities are more or less seen as belonging to the natural equilibrium. Thus, when we talk of environmental problems we implicitly mean that humankind is creating a problem in the life-sustaining system.

The consequence of this distinction is that it is important to establish the cause of a problem as either human-made or not. If humankind is not the cause of the problem, we are more inclined to adapt ourselves to the situation. But in a human-made problem there is a greater implication that we should act to remove its cause.

However, even minor changes can have large and structural effects (the Butterfly Effect, i.e. non-linear dynamics[3]). Thus, natural catastrophes

2 R.R. Rhew, B.R. Miller and R.F. Weiss, 'Natural Methyl Bromide and Methyl Chloride Emissions from Coastal Salt Marshes', *Nature* 403 (2000): 292-95.
3 Small changes in initial conditions can result in huge transformations. Cf: 'For want of a nail, the shoe was lost. For want of a shoe, the horse was lost. For want of a horse, the rider was lost. For want of a rider, a message was lost. For want of a message the battle was lost. For want of a battle, the kingdom was lost!'

could change the world for good. The natural phenomenon of an asteroid's collision with the Earth is supposed to be a main cause of the extinction of dinosaurs—definitely a lasting effect.

Neither is the distinction pedantic in political debates: discussion on human or natural causes of a problem occurs frequently. For example:

- What is the reason for declining male fertility rates?[4] Chemicals, lifestyle or blue jeans?

- Why are temperatures rising? Solar activity or greenhouse effect? See below

- What is the source of arsenic contamination of many of the tube wells in Bangladesh (see Table 2.2)?. Geological conditions, agricultural practices or something else?

Total number of districts in Bangladesh	64
Total area of Bangladesh	148,393 km^2
Total population of Bangladesh	120 million
Gross domestic product per capita (1998)	$260
WHO arsenic drinking water standard	0.01 ppm
Maximum permissible limit of arsenic in drinking water of Bangladesh	0.05 ppm
Number of districts surveyed for arsenic contamination	64
Number of districts having arsenic above maximum permissible limit	59
Area of affected 59 districts	126,134 km^2
Population at risk of the affected districts	75 million
Potentially exposed population	24 million
Number of patients suffering from arsenicosis	7,600
Total number of tube wells in Bangladesh	4 million
Total number of affected tube wells	1.12 million

TABLE 2.2 The problem of arsenic contamination of drinking water in Bangladesh

Source: www.eng-consult.com/arsenic/arsstat.html, 21 October 2005; © EngConsult Limited, Toronto, Canada

In 1991, two Danish scientists[5] claimed that there was a correlation between the Earth's average temperature and solar activity. Solar activity has generally increased over an interval of 100 years just as the Earth's temperature has done (Figure 2.3). Within the same interval, the Earth's average temperature has increased by approximately 0.7°C.

4   Cf. T. Colborn, D. Dumanoski and J.P. Meyers, *Our Stolen Future: Are We Threatening our Fertility, Intelligence and Survival? A Scientific Detective Story* (New York: Dutton, 1996).
5   Cf. E. Friis-Christensen and K. Lassen, 'Length of Solar Cycle: An Indicator of Solar Activity Closely Associated with Climate', *Science* 254 (1991): 698-700.

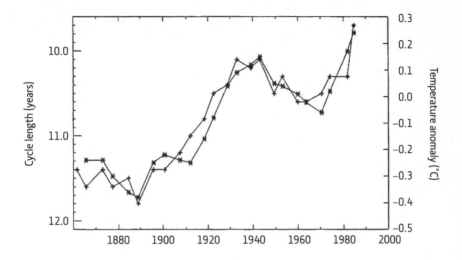

FIGURE 2.3 Correlation between solar activity and the Earth's average temperature

Reprinted with permission from E. Friis-Christensen and K. Lassen; © *Science* 254 (1991): 698-700

Even the finer structures in the two curves have similar appearances. Their paper triggered a lot of controversy and the last bend of the curve (around 1980) was later corrected. The general conclusion was that the affirmed correlation was less noticeable after 1980, and that therefore climate change could not be explained as being caused by solar activity alone.[6]

It has become increasingly difficult to establish whether humans or nature has caused specific catastrophes. Many 'natural' catastrophes are at least made worse by human action, e.g. deforestation, canalisation and urbanisation of river banks are often important factors when rivers flood. Deforestation also contributes to disasters such as droughts and erosion by allowing the soil to dry out more quickly. Droughts caused devastating fires in Indonesia and Brazil in 1997 and 1998. These fires occurred in tropical forests that are normally too moist to burn.

When is environmental change a problem?

If humans cause a change in the life-sustaining system, when do we recognise it as a problem? People are changing their natural environment constantly and by almost all of their activities. Which changes are seen as problems? Historically, a number of types of changes have been regarded as problems:

6   P.E. Demon and P. Laut, 'Pattern of Strange Errors Plagues Solar Activity and Terrestrial Climate Data', *Eos* 85.39 (2004): 370-74.

Disastrous flooding of Yangtze River

The costliest disaster of 1998 was the flooding of China's Yangtze River. It caused more than 4,000 deaths, dislocated 223 million people, inundated 25 million hectares of cropland and cost well over $36 billion. Heavy summer rains are common in southern and central China and flooding often occurs. But, in 1998, it became clear that other factors were at play.

In the past few decades, 85% of the forest cover in the Yangtze watershed has been cleared by logging and agriculture. Deforestation has left many steep hillsides bare. The loss of forests, which normally intercept rainfall and allow it to be absorbed by the soil, allowed water to rush across the land carrying valuable topsoil with it. In addition, the river's natural flood controls had been undermined by numerous dams and levies, and a large proportion of the basin's wetlands, which usually act as natural 'sponges', had been filled in or destroyed. All these changes reduced the capacity of the Yangtze basin to absorb rain and greatly increased the speed and severity of the resulting run-off.[7]

- Changes that threaten our health
- Changes that threaten species or ecosystems
- Changes that affect the economic benefits we receive from natural processes
- Changes that affect the benefits that others (future generations, under-developed countries) receive from natural processes

These four types of changes are considered below.

Threats to health

Historically, the first environmental problems emerged as threats to public health. People did not categorise specific problems as environmental problems but, with hindsight, various rulings in medieval cities might be recognised as environmental regulation. Most of these regulations were related to waste disposal. The city of Delft in the Netherlands, for example, had a special windmill installed to improve the circulation of the waters in its canals. The canals were used as an open sewer, as well as a fresh water supply for the city and its breweries. In 1588, Simon Stevin studied this mill and improved it—an early example of environmental innovation.[8]

7   J.N. Abramovitz, 'Natural Disasters: At the Hand of God or Man?' (23 June 1999); www.enn.com/arch.html?id=15820, 21 October 2005.
8   K.F. Mulder, 'De Hollandse Molen: Monument of werktuig?' (Boerderijcahier 8701; Enschede: University of Twente, 1987).

Horse manure, urine and horse carcasses left in the streets of cities were considered important (environmental) problems in the 19th century and contributed to the spreading of various diseases. Although the 20,000 New York victims from manure-related diseases mentioned by Bolce in 1908 could be an exaggeration, we tend to underestimate the horse manure problem now.[9] Indeed, automobiles were considered the solution to a serious environmental problem.

For most people today, the threats to human health are still the main reason to worry about the environment. Damage to human health by smog, pesticide residues, carcinogenic compounds and the contamination of drinking water are important issues of concern.

The effects of pollution on public health are still largely unknown. About 100,000 synthetic chemicals are currently being manufactured and only a small fraction of them has been tested thoroughly. New facts are frequently discovered. About ten years ago, for example, the relation between chemical pollution and human fertility was brought to our attention.[10]

Tributyltin (TNT) was introduced as a non-polluting anti-fouling agent for ships until it was discovered that the compound disrupts the endocrine system of various animals. Recently, it was discovered that tiny dust particles (from smog) can zoom through human lungs up to two times faster and penetrate deeper than previously assumed.[11]

The direct effects of pollution on human health do not occur very frequently in developed countries. Health effects in industrial societies more often occur due to prolonged exposure. Important public health disasters have also been caused by major accidents.

There are many delayed effects from various compounds that we use. Well-known examples include:

- The debate on the long-term effects of smoking
- Noise, which might cause hearing damage[12]
- Dust, which might lead to lung diseases[13]
- Organic solvents may cause Organic Psycho Syndrome (damage to the central nervous system)[14] and probably decrease male fertility

9 J. Tarr and C. McShane, 'The Centrality of the Horse to the Nineteenth-Century American City', in R. Mohl (ed.), *The Making of Urban America* (New York: SR Publishers, 1997): 105-30; www.enviroliteracy.org/article.php/578.html, 21 October 2005.
10 T. Colborn, D. Dumanoski and J.P. Meyers, *Our Stolen Future: Are We Threatening our Fertility, Intelligence and Survival? A Scientific Detective Story* (New York: Dutton, 1996).
11 G. Pinholster, 'Smog Impacts: Hurtling through airways, tiny particles may do more damage than previously assumed', EurekAlert, 1999; www.eurekalert.org/pub_releases/1999-08/UoD-SiHt-160899.php, 21 October 2005.
12 Cf. www.mckinley.uiuc.edu/Handouts/noiseeh/noiseeh.html, 21 October 2005.
13 Cf. www.agius.com/hew/resource/lung.htm, 21 October 2005.
14 Cf. members.chello.nl/~w.vanmidlum/engels.html, 17 March 2006.

Porto Marghera, Venice

Since 1951, Enichem and Montedison have produced chlorine, vinyl chloride monomer (VCM) and polyvinyl chloride (PVC) near the Venice lagoon. In 1994, labourer Gabriele Bortolozzo handed a report to the authorities containing evidence of tens of casualties related to toxic releases. Based on this evidence, the first Italian criminal lawsuit for toxic emissions began.[15]

The Bhopal (India) accident

On 3 December 1984, methyl isocyanate gas leaked from a tank at Union Carbide's chemical plant in Bhopal, India. It led to an enormous tragedy: 3,800 people died, 40 suffered from permanent total disability and 2,680 suffered from permanent partial disability as a result of the toxic release.[16]

Lead

Some researchers believe that behavioural and neurological defects exhibited by most of the emperors of Rome between AD 15 and 225 were caused by lead poisoning. Romans used lead for water ducts, utensils, ornaments, wine storage, coins and jewellery.[17]

Tetraethyl lead was generally used as fuel additive until 1975[18] and gradually phased out until 1995. Today, despite known health risks, lead remains among the most widely used of metals.[19]

15 Cf. see the documentary by Paolo Bonaldi entitled 'Porto Marghera, Venice: The Story of a Lethal Deception', Festival International de Cinema del Medi Ambient; www.mediapro.es/eng/documentales.htm#6, 21 October 2005.
16 Cf. P. Shrivastava, *Bhopal: Anatomy of a Crisis* (London: Paul Chapman Publishing, 2nd edn, 1992). See also www.bhopal.com, 21 October 2005.
17 D.W. Green, 'The Saturnine Curse: A History of Lead Poisoning', *Southern Medical Journal* 78.1 (1985): 48-51.
18 J.L. Kitman, 'The Secret History of Lead', *The Nation*, 2000; www.mindfully.org/ Pesticide/Lead-History.htm, 21 October 2005.
19 See www.epa.gov/iaq/lead.html, 21 October 2005.

- Mercury poisoning
- Asbestos may cause cancer even decades after exposure[20]
- Lead poisoning

Threats to ecosystems or wildlife

Most views regarding sustainable development (including the one implicitly expressed in Chapter 1) take an **anthropocentric** (human-centred) position, i.e. problems have negative consequences for humans.

However, another conceptualisation of problems, as creating a threat to any living creature or ecosystem, could be observed by the end of the 19th century: the **conservationists** headed by Gifford Pinchot who founded Yellowstone Park in the USA in 1872 wanted to conserve natural resources for later generations. Gifford's assistant John Muir wanted to preserve nature itself, not just specific species or locations. The two men fell out and John Muir became an important **preservationist**. He founded Yosemite Park in 1892. The basis of preservationist thinking is the recognition that nature has its own intrinsic value. Human needs were unimportant—at least not the key issue—for preservationists. The destruction of nature itself was the problem.

In Europe, preservationism came later than in the USA. The Dutch Society for Preservation of Natural Monuments[21] purchased its first area in 1906. In Germany, the Society for Preservation of the Lüneburg Heath[22] was founded in 1909. In Spain, two National Parks were formed in 1918.[23] In the UK, the Standing Committee on National Parks began life in 1936.[24]

Humans were often portrayed as stewards over nature. Such a view was related to (interpretations of) Christianity in which humankind received the divine task of stewarding God's creation. In other religions, similar lines of reasoning are often present:

> Even things which appear to you to be superfluous in the world, such as flies, fleas and mosquitoes, they are also part of the creation of the world, and God performs His operations through the agency of all of them, even through a snake, mosquito or frog.[25]

---

20 Cf: www.asbestos.org/asbestos/test_frames.html, 21 October 2005.
21 Vereniging tot Behoud van Natuurmonumenten.
22 Verein Naturschutzpark Lüneburger Heide.
23 F. Navarro, *Parques y espacios naturales protegidos naturales de España* (Barcelona: Salvat, 1997).
24 See www.cnp.org.uk/50th_anniversary.htm, 21 October 2005.
25 Torah, Midrash Bereishis Rabbah 10:7.

Nowadays, an economic argument is often added to this reasoning: biodiversity is an economic value because the variety of species is a genetic resource for plant and animal breeding. However, this is only ever an additional argument for preservationists—never *the* argument.

The preservationist view conflicts with industrial development only when industry is directly threatening nature reserve areas or the survival of species. There is no general conflict of interests. People who had earned their money from economic activities that were far from environmentally benign have sometimes played major roles in preservationism.

Threats to economic benefits

The third conceptualisation of problems is a more or less utilitarian vision. Industrial society leads to such growth that nature is over-exploited. In many cases, this has created risks for public health and wildlife. As a result of the enormous growth in population and production, however, a new factor became important. Natural resources became scarce.

In history, scarcity was often at the root of ingenuity. In 17th-century England, scarcity of firewood led to coal mining. The scarcity of coal in near-surface layers in turn led to the development of pumping equipment and steam engines. Scarcity also led to trade.

In our times, scarcity is not just a local phenomenon. For various raw materials, we can observe a threat of global scarcity. The world is finite, and therefore trade can no longer help us. We have only our ingenuity to search for solutions.

Overfishing was one of the first economic problems of this kind. Harvesting from biological resources is renewable as long as the speed of harvesting does not exceed the speed of natural recovery of the ecosystem. If it does, stocks will decline. Very often, the reaction is to 'improve' harvesting methods to keep the yield of the harvest intact. Therefore, the speed of decline will increase. If the process is not stopped in time, irreversible damage is caused. Fisheries for Californian sardines and sardine processing and packing (portrayed in John Steinbeck's famous novel *Cannery Row*[26]) disappeared in the early 1950s.[27] In the 1936/37 season, three-quarters of a million tons of sardines were taken from Californian waters. In subsequent years, the fleet was enlarged as the catch per boat began to drop. But the fishing fleet ignored advice from biologists. By 1957/58, the catch was only 17 tons.[28] Only recently have there been signs of a recovery. As in many other cases, it is

26 J. Steinbeck, *Cannery Row* (London: Heinemann, 1945).
27 J. Radovich, 'The Collapse of the California Sardine Industry: What Have We Learned?', in H.M. Glantz and J.D. Thompson (eds.), *Resource Management and Environmental Uncertainty* (New York: John Wiley Interscience, 1981): 107-36.
28 darwin.bio.uci.edu/~sustain/bio65/leco4/b65leco4.htm#FISH AND FISHERIES, 21 October 2005.

debatable if sardines disappeared only as a result of human action. Climate change might also have played a role.[29]

In 1992, 40,000 Canadian fishermen lost their jobs when cod fishing at the Great Banks off the east coast of Newfoundland collapsed.[30]

Threats to others' benefits

Our use of the environment may not have serious consequences for us, but it might limit the options of others to use the environment. As the amount of natural resources is limited, who has the right to utilise them? Can it be justified that there will be no natural supplies of crude oil for our great-grandchildren? Is it justifiable that about 20% of the world's population uses 80% of the world's resources? If industrialised countries are responsible for the greenhouse effect, shouldn't they pay damages to African peasants who are forced to move, to Bangladesh that will be flooded more often, or to coral islands that will vanish? Should less-developed countries be allowed to increase their emissions to enable them to create industrial capacity?

In 1992 at the UN Conference on Environment and Development, the international community acknowledged the legitimacy of the demand by under-developed countries to have a larger share of the wealth produced in the world.

Causes of problems in the life-sustaining system

Problems in the life-sustaining system can be caused by various mechanisms. A well-known sub-division is as follows:

- **Pollution:** those problems that originate from the addition of chemical substances or physical phenomena to our environment. Environmental pollution problems can in principle be solved by removing the source of pollution and cleaning up the affected area

- **Exhaustion:** those problems that originate from the removal of resources from nature. In case of exhaustion of plants or wildlife, natural processes can be restored if exhaustion has not led to the extinction of species. Exhaustion of fossil fuels, minerals and ores is generally an irreversible process. It implies that the world just has to live without the natural reserves of those substances

- **Degradation of ecosystem.** This means losses in the quality of ecosystems. These problems might be caused by various changes such as road construction and other public works that create barriers for species migration or the recreational use of ecosystems that disturb animals

29 whyfiles.org/139overfishing/2.html, 21 October 2005.
30 arcticcircle.uconn.edu/NatResources/cod/mckay.html, 21 October 2005.

- **Environmental risks.** These are problems that will not occur under normal circumstances, but which are devastating if they do. The ESA/NASA Cassini–Huygens mission was a typical example of an environmental risk. This spacecraft carried about 33 kg of plutonium dioxide as its source of power. It was launched in October 1997 and came perilously close to Earth in August 1999.[31] If it had hit the planet, the effects would have been devastating

- **Social conflicts.** As environmental problems are closely connected to societal interests, social conflicts can cause environmental problems. The utilisation of the environment has often led to war (possession of raw materials, water wells, good land and colonies). Moreover, war is generally destructive for humans and their natural environment

Hazards of war

Agent Orange was the code name for a herbicide developed for military use in tropical climates. Its purpose was to deny cover and concealment to an enemy in dense terrain by defoliating trees and shrubbery. Agent Orange was principally effective against broadleaf foliage such as the dense jungle-like terrain found in South-East Asia. The product was brought into ever-widening use during the height of the Vietnam War (1967/68), though its use diminished and eventually ceased in 1971.

The earliest health concerns about Agent Orange were about its contamination with dioxin. This causes a wide variety of diseases, many of them fatal.

Another environmental disaster came about due to the Gulf War in 1991 when Iraqi troops set fire to hundreds of Kuwait oil wells (see Figure 2.4). The plume of smoke could be observed into the Indian subcontinent.[32]

Although various causes might lead to war, structural injustice such as the enormous gap in affluence between rich and poor countries may contribute to conflict. The conflicts caused by this problem are generally not between rich and poor countries, but between the poor and the rich minorities that often form the ruling class in these nations. Developed countries should therefore give overriding priority to the needs of the poor in developing countries.

31 See saturn.jpl.nasa.gov/spacecraft/safety.cfm, 21 October 2005.
32 See special issue of *Journal of Geophysical Research* 97.D13, 20 September 1992.

FIGURE 2.4 **Fires in Kuwait**

Source: www.redadair.com/thriller.html, 25 October 2005; © Adair Enterprises, 1999

Spatial distribution of effects of problem

Environmental problems vary enormously in space and time. The scope of the areas affected is not just important as a measure of the severity of the problem. If effects occur in places that are far removed from their cause, the conflict of interests is harder to solve.

An example of this occurred in the 1960s and 1970s when market gardeners in the Westland near Rotterdam could no longer use water from the River Rhine for irrigation. This was because potassium mines in eastern France were discharging large amounts of salts into the river, which made the water unfit for irrigation.

Another example was Spain's National Hydrological Plan.[33] This aimed to transport vast quantities of freshwater from the Ebro River to the Malaga region in order to stimulate tourism and agriculture in southern Spain. However, this would have endangered the natural preserve areas and agriculture in the Ebro Delta, as salinity would have increased sharply.[34]

A virtual conflict might arise if the time dimension is at stake as people might feel that future generations will be saddled with our waste. This time dimension was especially important in the debates on the acceptability of nuclear power plants, as the waste generated (mixtures of various radioactive elements) would remain harmful (radiation, toxicity, danger of terrorist use) for thousands of years.

33 The plan was discontinued after Spain's 2004 elections.
34 www.rivernet.org/Iberian/planhydro.htm, 21 October 2005.

Space effects are present in many modern debates on sustainable development. The issue is especially focused on intergenerational equity, i.e. is it justified that our generation consumes all the world's resources and leaves the pollution to our (grand-)children? It is often stated that, in a sustainable society, we should consume only what can be replenished at the same rate. However, will future generations need the same quantities and qualities of resources as we do? One could also argue that, if we use too much of a specific resource, we should at least try to develop improved technology that would allow future generations to fulfil their needs without that resource.

Chernobyl incident

In 1986, a series of human errors brought about an accident at a nuclear plant in Chernobyl in the Ukraine. This led to an explosion and radioactive contamination, killing 31 people immediately and hospitalising another 150 with severe radiation sickness. Approximately 130,000 people from a 30 km² zone around the atomic plant were evacuated. Many thousands more are expected to die prematurely because of their exposure to radiation released from the stricken plant.[35]

While the plume was detectable in the northern hemisphere as far away as Japan and North America, countries outside Europe received very little deposition of radionuclides from the accident. No deposition was detected in the southern hemisphere.

In fact, the weather was the cause of an uneven deposition of radionuclides. For example, the estimated average depositions of caesium-137 in the provinces of Upper Austria, Salzburg and Carinthia in Austria were 59, 46 and 33 kBq/m² respectively, whereas the average caesium-137 deposition in Portugal was 0.02 kBq/m².[36]

Spatial environmental effects are often divided into:

- **Global:** greenhouse effect, dilution of ozone layer, decreasing biodiversity, exhaustion of resources, overpopulation, nuclear fallout, etc.

- **Continental**: a typical continental problem is the atmospheric transport of pollutants such as sulphur oxides (SO_x) and nitrogen oxides (NO_x) (most pollutants are washed away by the time they have crossed an ocean)

35 For an overview with BBC News clips, see www.chernobyl.co.uk, 21 October 2005.
36 Nuclear Energy Agency (NEA), 'The Release, Dispersion and Deposition of Radionuclides', in Chernobyl: Assessment of Radiological and Health Impact 2002. Update of Chernobyl: Ten Years On (2002): ch. 2; www.nea.fr/html/rp/chernobyl/co2.html, 21 October 2005.

- **Regional:** e.g. the pollution of an estuary or the noise of an airport

- **Local:** e.g. soil pollution, various nuisances from garbage dumps, odour, etc.

- **Indoor:** e.g. air pollution such as smoke or radiation such as radioactive radon release from concrete

For some problems, other spatial scales are more suitable:

- **River basins.** So-called fluvial environmental problems generally stem from water pollution (which might be chemical but also thermal, i.e. power stations using cooling water) or excessive water use in relation to its availability (e.g. excessive use of irrigation water in central Asia for cotton growing). Other river problems include river bottom pollution, flooding caused by canalisation, river eco-system disturbance (e.g. by hydropower barrier dams) and eco-disasters by accidents

- **Coastal seas** are often quite separate ecosystems. Pollution (by coastal states) is often only slowly diluted. Various countries may compete for the fish. Tourism may be affected by pollution. Control of the merchant marine may also be important to prevent oil spills or large eco-catastrophes caused by oil tankers

The offshore industry may also create serious environmental problems. These problems can generally only be solved by co-operation between coastal states such as those lining the Mediterranean.

One could argue that the human body is the spatial scale for some environmental problems, though we would usually call these medical rather than environmental problems. Such a problem might be the severe contamination of breast milk with human-made pollutants. The WWF recently stated that more than 350 human-made pollutants had been identified in the breast milk of women in the UK. These include 87 dioxins.[37]

Gravity of problems

In order to set policy priorities, it is necessary to somehow determine the gravity of various environmental problems. Exposure effect relations are often required for the ultimate assessment of the environmental effects of various environmentally harmful actions. These relations may show a No Observed Effect Level (NOEL). In some cases, decision-makers might use NOEL to set standards. However, there are many cases where every exposure increases the effect and a NOEL is not present. Moreover, if long-term effects are taken into consideration, it is often very hard to show that a NOEL really applies.

37 news.bbc.co.uk/low/english/health/newsid_380000/380948.stm, 21 October 2005.

Mediterranean Action Plan

The Mediterranean Action Plan (MAP) strives to protect the environment and to foster development in the Mediterranean Basin. Under the auspices of the United Nations Environment Programme (UNEP), 16 Mediterranean states and the EU adopted the MAP in Barcelona, Spain, in 1975. Its legal framework comprises the Barcelona Convention adopted in 1976 and revised in 1995, and six Protocols covering specific aspects of environmental protection. The MAP has served as the basis for the development of a comprehensive environment and development programme in the region. It covers coastal zone management, pollution assessment and control, protection of ecosystems and preservation of biodiversity. In 1995, it was revised to become more action-oriented and an instrument for sustainable development in the region.[38]

An alternative route is therefore to determine an Accepted Effect Level and to determine the corresponding exposure level. This route is not easy. How to reach consensus on the effect level that we are prepared to accept? In general, the effects that we accept depend on our attitude to the cause of the effects. Citizens are often more willing to accept nuisance from activities they support when they consider climate change or the future of energy supplies.[39]

In practice, environmental standard setting is a political process that is determined by the usual factors in public policy-making (i.e. power, group interests and media attention). However, one can reflect on some general principles for environmental policy:

- Is the problem lethal for humans?
- Is the problem lethal for (large) ecosystems?
- Will the effects remain for long once the cause of the problem is removed?
- What is the spatial magnitude of the effects?

In general, there are no straightforward answers to these four questions. For instance, in epidemiological studies, problems that seem just a nuisance (odour, noise) may turn out to cause premature death. Degradation of ecosystems is often completely unexpected and may be simply because some minuscule element is affected. Deteriorating ecosystems might also affect human life as they could lead to starvation, etc. The third and fourth ques-

38 See www.unepmap.org, 21 October 2005.
39 For example, Maarten Wolsink found that citizens worried about the future of electricity supplies accepted the nuisance from wind turbines (*Maatschappelijke acceptatie van grote windturbines* [The Hague: VROM, 1987]).

tions are related to equity issues. Are many of the victims others than those who caused the problem?

An assessment of the gravity of environmental problems must be based on some basic assumptions regarding the weights that we attribute to:

- The 'health' of ecosystems
- The physical conditions under which we live
- Solidarity with the poor and with future generations

Precautionary principle

The effects of environmental degradation are often hard to establish scientifically. Because some problems could be devastating in their effects, we cannot afford the time to wait until scientific consensus is established on the facts. For example, after half a century of scientific research there is still no absolute scientific proof of climate change and its causes.[40] Can we afford to postpone measures?

For this reason, the United Nations Conference on Environment and Development at Rio de Janeiro in June 1992 adopted the **precautionary principle**:

> In order to protect the environment, the precautionary approach shall be widely applied by States according to their capabilities. Where there are threats of serious or irreversible damage, lack of full scientific certainty shall not be used as a reason for postponing cost-effective measures to prevent environmental degradation (Principle 15 of UN Declaration of Rio de Janeiro).[41]

International adoption of this principle will not solve the dilemma between taking measures and searching for further evidence in concrete cases. There will always remain scope for discussion regarding the seriousness of the threat and the scientific uncertainties.

---

40 S.R. Weart, *The Discovery of Global Warming* (Cambridge, MA: Harvard University Press, 2003).
41 www.unep.org/Documents.multilingual/Default.asp?DocumentID=78&ArticleID= 1163, 21 October 2005.

The main threats that humankind faces today

The problems of the life-sustaining system can be compartmentalised into:

- Biosphere (all living material)
- Hydrosphere (oceans, lakes and rivers)
- Atmosphere (the air surrounding the Earth)
- Lithosphere (the Earth's crust)

An extra compartment might be **outer space**. For example, space organisations are increasingly hampered in their activities by space debris—objects of various sizes that are left in space and remain in orbit. Although this division into compartments will suffice for our purposes, the compartments are often strongly interconnected. For example, polluted soil might contaminate water and air, while polluted water could create polluted sediments.

Biosphere

Ecosystems can be threatened by:

- **Pollution.** Various chemicals have direct toxic effects. However, longer-term toxic effects and effects on reproduction are just as harmful, though less visible. The pesticide DDT and organochlorine compounds such as polychlorinated biphenyls (PCBs) hardly break down at all in the natural environment and have long-ranging effects on living organisms. These long-term effects are only partly known

- **Extirpation or decline of a species.** Hunting and fisheries not only drive certain species towards extinction (e.g. whales in oceans, herring in the North Sea, elephants on the African plains, otters or the dodo[42]), the whole ecosystem of which these creatures are part of is affected. Open plains often disappear if there are no cattle to graze them. Inland wetland ecosystems may disappear completely if beavers are wiped out. In the case of fast-replicating species such as microorganisms, near extirpation can lead to survivors that are able to withstand the threat affecting their numbers; but the result may be more dangerous species of microorganisms

- **Increase in naturally occurring elements.** Manure is a natural element in the ecosystem but, if too much is produced, the ecosystem is seriously affected. In particular, emissions from the bio-

---

42 A flightless bird from Mauritius, which was extirpated 80 years after its first sighting in 1600. See www.amnh.org/exhibitions/expeditions/treasure_fossil/ Treasures/Dodo/dodo.html?dinos, 21 October 2005.

industry (pigs, chickens, etc.) can seriously harm local ecosystems. An over-abundance of animals may also harm ecosystems. Packing animals closely together allows contagious diseases to spread more rapidly. This is also true for humans. Nowadays we are more densely packed and move around much more and much faster. Epidemics today can spread round the world in a few days, whereas the Black Death took several years to reach all corners of a continent. It began in the Black Sea region in 1347 and reached England, Scandinavia and Portugal in 1349, but Estonia only in 1352.[43] Too high a concentration of nutrients in an ecosystem is called eutrophication. In aquatic environments, this can lead to algae blooms that reduce the oxygen content of water and kill fish. Eutrophication of soil can radically change plant life[44]

- **Introduction of non-indigenous species** can have dramatic effects on local ecosystems. Examples include:
  - The *introduction of rabbits* in Australia in 1859 when 24 wild rabbits were released in South Australia. Controlling the numbers once they had invaded the land was almost impossible and today there are about 1 billion rabbits in Australia. They compete for food with indigenous species, which become endangered. The rabbits also destroy plants, thus causing the land to erode. Rabbits cause over €100 million of damage each year to the Australian economy[45]
  - The *migration of aquatic species via ships and canals*. Many species from the Red Sea have made their way to the Mediterranean via the Suez Canal since it opened in 1870.[46] Migration of aquatic life by canals is also occurring by the Rhine–Main–Danube Canal which was opened in 1992. Species from the Black Sea region are colonising Northern Europe. Empty ships that take ballast in their tanks transport aquatic life. The problem is enormous. In the North American Great Lakes more than 150 non-indigenous species have completely transformed the original ecosystem
  - The *introduction of exotic plants*. For example in South Africa, imported plants threaten indigenous species, use more water (thus leading to drought) resulting in more bush fires[47]

43 See www.insecta-inspecta.com/fleas/bdeath, 21 October 2005.
44 Cf. the website of the Sea Grant Nonindigenous Species (SGNIS); www.sgnis.org, 21 October 2005.
45 'History of Rabbits in Australia', library.thinkquest.org/03oct/00128/en/rabbits/history.htm, 17 March 2006.
46 'CIESM Atlas of Exotic Species in the Mediterranean', www.ciesm.org/online/atlas/index.htm, 17 March 2006.
47 H. Moleman, 'Slegs vir eie plante', *De Volkskrant*, 27 May 2000: 5W.

- **Biodiversity** is an important asset for our planet:
  - It helps us overcome disasters as there are often mutants within a species that can resist diseases
  - It is a resource for finding new treatments for diseases
  - It may help us understand the processes of life as it is the product of millions of years of evolution
  - It is the rich heritage of our planet

  The main reason for many species becoming extinct or threatened is the destruction of ecosystems. However, there is a paradox: although biodiversity is decreasing, the number of species at a local level has often increased. For example Hawaii has a total of 2,690 plant species, 946 of which are non-indigenous.[48] Of the native species, about 800 are currently endangered.[49] The paradox between local increase and global decrease of biodiversity is due to the introduction of non-indigenous species. Our mixing of species is making the world locally more varied but globally more uniform

- **Biotechnological risks.** Biotechnology is increasingly able to produce genetically modified organisms. This could create a risk as it is not possible to know all the effects in advance. Some of the organisms may threaten (elements of) ecosystems. Optimisation of agricultural crops by genetic modification could narrow the genetic variety of these plants

Hydrosphere

The total amount of water on Earth is about 1 million km^3, of which 97% is in the oceans. The rest is freshwater or glaciers.

Most freshwater can be found in deep subterranean reservoirs. The water in the oceans is moving constantly. At the surface, the ocean currents follow the dominant winds. In the deep of the ocean, the flow is often reversed. The oceans transport heat from the Equator to higher latitudes and have a considerable influence on regional climates. For example, the Gulf Stream transports heat from the Atlantic near Florida to northern Europe.

The water moves in cycles: it evaporates at the ocean surface, condenses into clouds, is transported to continents, precipitates onto the land, where some of it is used by organisms, but the major part is transported back to the ocean by rivers.

48 L.G. Eldredge and S.E. Miller, 'Numbers of Hawaiian Species: Supplement 2, including a Review of Freshwater Invertebrates', *Bishop Museum Occasional Papers* 48 (1997): 3-32.
49 P.M. Vitousek, 'Diversity and Biological Invasions of Oceanic Islands', in E.O. Wilson and F.M. Peter (eds.), *Biodiversity* (Washington, DC: National Academy of Sciences, 1988): 181-89.

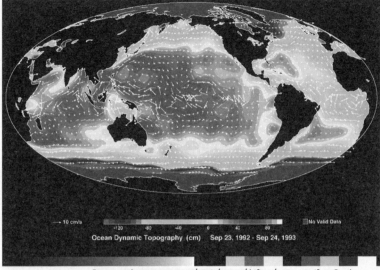

This water cycle is important for it supplies us with freshwater. It also cleans the atmosphere as aerosols (airborne particles), SO_x and NO_x, are washed out by rain. SO_x and NO_x cause the rain to become 'acid'. Acid rain has serious negative effects on surface waters, which acquire a low pH level. This affects aquatic ecosystems. Forests also suffer, especially at high elevations. The peaks of the Krkonose, the Czech Republic, are an extreme example. These peaks are often shrouded by fog, which is especially acid due to the proximity of a power plant with huge acid emissions in nearby Poland. Marble and limestone are also affected and so details of the façades of old monuments become eroded. Car paints can show etched areas.

Oceans

Oceans are polluted in a number of ways:

- **Oil spills and leakages.** The accidental spills of large quantities of oil such as those involving the *Torrey Canyon* (UK, 1967), *Amoco Cadiz* (France, 1978), *Exxon Valdez* (Alaska, 1989), *Erika* (France, 1999) and *Prestige* (Spain, 2002) receive widespread media coverage. However, operational oil pollution could be more important than accidental spills. Accidental spills amount on average to an estimated 150,000 tons per year while operational discharges of oil could be much larger.[50] However, the deposition of oil into the oceans due to

50 Global Marine Oil Pollution Information Gateway; oils.gpa.unep.org/facts/operational.htm, 21 October 2005.

FIGURE 2.6 Trees killed by acid rain in Krkonose, Czech Republic

Source: gei.aerobatics.ws/czech_republic.html, 21 October 2005; © Günther Eichhorn

unburned fuel that is released in the atmosphere and washed into the sea and to the oil content of rivers and gutters is probably far larger.[51] Oil and chemicals can also become ticking time-bombs if the vessel sinks but the tanks, canisters and containers remain intact. The effect is the same as dumping highly toxic and nuclear waste into the ocean, as there will come a time when the packaging corrodes away

- Ocean dumping of **nuclear waste** has been banned since 1993, but operational discharges of low radioactive fluids continue

- **Litter** is another problem for oceans. Plastics break down very slowly. Plastic litter can be found in every corner of the oceans— even in the stomachs of seabirds. Large remains of plastic nets can kill sea mammals long after the nets were used for fishing

- **Anti-fouling paints** are used on ships to prevent microorganisms growing on their hulls. TNT affects the endocrine systems of vertebrates but has been subject to international regulation since 2003. There are many other harmful contaminants: the use of PCBs and DDT has been controlled for some time but these chemicals can still be found in marine life

- Marine pollution may change in **contact with sediments**. From 1932 to 1968, an estimated 27 tons of mercury compounds were dumped into Minamata Bay, Japan. The mercury was methylated in the sediments of the bay. In the 1950s, thousands of people whose

51 www4.nas.edu/onpi/webextra.nsf/web/oil?OpenDocument, 21 October 2005.

Amoco Cadiz

On 16 March 1978, the supertanker *Amoco Cadiz* filled with 223,000 tons of crude oil lost its entire cargo in the Atlantic Ocean off Portsall, Brittany, covering more than 130 beaches in oil up to a depth of 30 cm. The oil slick from this immense oil spill, which was almost eight times the size of the 1989 *Exxon Valdez* spill off the coast of Alaska, caused an enormous environmental disaster. Over 30,000 seabirds died, along with 230,000 tons of crabs, lobsters and other fish. The area's prized oyster and seaweed beds, which provided income for many local inhabitants, were completely destroyed. The accident occurred during a storm when the ship lost its ability to steer and ran aground on rocks. As a tug attempted to tow the tanker out of trouble, the ship broke up.[52]

normal diet included shellfish from the bay developed symptoms of methylmercury poisoning, which became known as 'Minamata Disease'[53]

Inland water

Inland water is essential for humans: we drink it, we use it to grow food, we use it in production, we use it to remove our waste and we use it for recreation. Our consumption of water creates many problems, especially in densely populated areas: local droughts can be created and upstream pollution can harm downstream consumption considerably. The functions of inland water often cannot be combined and must be carefully planned.

The problems that arise from lack of planning of inland water resources are illustrated by the Aral Sea. The Soviet Union diverted the Amu and Syr waters for cotton irrigation and the construction of the Kara Kum canal in 1954. This led to a severe decline in the Aral Sea. Since 1960, its mean level has dropped by 10 m and its area has almost halved, resulting in an increase in water salinity and more extreme winter and summer temperatures. Since 1990, the Aral Sea has split into two: a large area to the south and a much smaller area to the north.[54]

This problem is not unique. In various dryer parts of the world, the use of river water causes serious problems. Armed conflict has sometimes arisen over irrigation water: e.g. water from the River Jordan (Israel–Jordan) and the River Tigris (Turkey–Iraq).

---

52 'Twenty Years After the Amoco-Cadiz', International Scientific Symposium held in Brest, France, 15–17 October 1998; www.mairie-brest.fr/amoco-symposium/index.html, 21 October 2005.

53 See www.nimd.go.jp/english/index.htm, 21 October 2005.

54 wwwcpg.mssl.ucl.ac.uk/orgs/cp/html/hydro/Aral.html, 21 October 2005.

Inland water is sometimes severely polluted. Inland fisheries have disappeared from many inland waters due to a lack of fish or because the fish had become inedible.

Accidents can also play a role. In November 1986, a fire at the Sandoz chemical factory in Basle, Switzerland, caused a major leak of contaminated water into the River Rhine. The factory had contained 840 tons of pesticides, fungicides, dyes and other toxic chemicals. These chemicals mixed with the large quantities of water that were poured onto the flames. A resulting 30 tons of highly toxic waste entered the waters of the Rhine and half a million fish were killed instantly. The spill also caused a major drinking water crisis on the Rhine banks as the intake of river water had to be cut off.[55]

Climate change

The issue of overriding importance for the atmosphere is climate change. This has consequences for all other spheres.

The Earth's energy balance

The sun heats the Earth. The atmosphere regulates the Earth's surface temperature. Solar heat is transmitted mainly through high-frequency radiation—visible light. This radiation penetrates the atmosphere without heating it and heats the Earth's surface. The warm surface emits heat rays with much lower frequencies (infrared). The atmosphere is not completely permeable to this infrared radiation as CO_2, water vapour and methane, especially, absorb it. This heat is in part returned to the surface. Hence, greenhouse gases act as a blanket, preventing the Earth from cooling down.

The absorption of infrared radiation by gases in the atmosphere is called the greenhouse effect. The greenhouse effect is crucial. Without greenhouse gases, the afternoon temperature at surface level could rise to 82°C (the less important infrared from the sun would also reach the surface). During the night, temperatures would drop to –140°C. The average temperature on the surface of the Earth would be –17°C instead of +15°C now. The growing ice and snow mass would reflect more sunlight and therefore contribute to lower temperatures, making the Earth uninhabitable. Thus the greenhouse effect is crucial for our survival. Figure 2.7 summarises its principle.

However, the greenhouse effect is also a problem. The rising levels of CO_2 in the atmosphere—caused by the burning of fossil fuels—reinforce the greenhouse effect. Over the past 150 years, the concentration of CO_2 in the atmosphere has risen from 290 parts per million (ppm) to 345 ppm. Various other gases have similar effects. Although CO_2 is the most common greenhouse gas in the atmosphere, methane (CH_4), nitrous oxide ('laughing gas') (N_2O), water vapour and various hydrocarbons are much more potent greenhouse gases.

55 www.edict.com.hk/vlc/textview/texts/sandoz.htm, 21 October 2005.

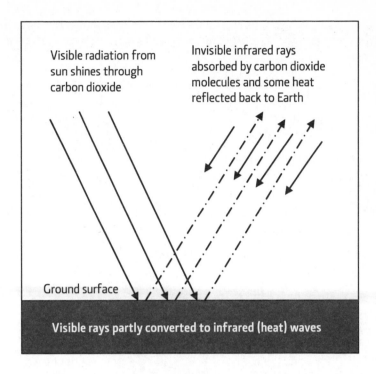

Figure 2.7 Principle of the greenhouse effect

Methane enters the atmosphere from leaking gas pipes, rotting organic material, animal breeding and the wet cultivation of rice. As methane is a valuable resource, preventing its release could possibly be environmentally effective and profitable.

According to the Intergovernmental Panel on Climate Change (IPCC), the average increase in surface temperatures over the 20th century was $0.6°C\pm0.2°C$.[56] A further increase ranging from 1.4 to 5.8°C is predicted by 2100 and the global average sea level will rise 0.09–0.88 m. Climate change will also affect other climate parameters such as rainfall, winds and the occurrence of extreme weather.

Part of climate change is probably due to a natural phenomenon, i.e. increased solar activity. But anthropogenic climate change is still evident when this effect is taken into account.

56 'Climate Change 2001: Summary for Policymakers': 5; www.ipcc.ch/pub/un/ syreng/spm.pdf, 17 March 2006.

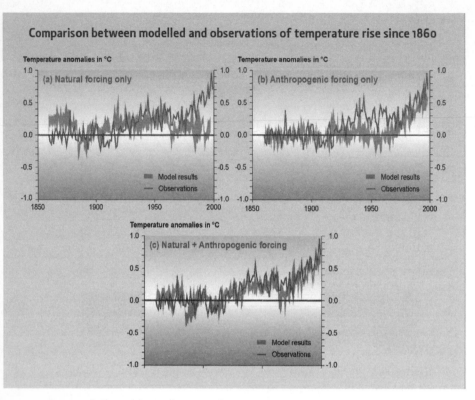

Comparison between modelled and observations of temperature rise since 1860

FIGURE 2.8 Temperature change and its causes

Source: www.ipcc.ch/present/graphics.htm, 21 October 2005

Uncertainties

The margins in these estimates are caused particularly by the potential occurrence of **positive** and **negative feedback loops**. Rising temperatures may create effects that in turn counteract temperature rises (negative feedback loop, as the effect counteracts itself). Rising temperatures may also create effects that lead to higher temperatures (positive feedback loop, as the effect reinforces itself). Examples of negative feedback loops include:

- Faster growth of vegetation caused by the higher CO_2 levels in the atmosphere. Plants grow faster in higher CO_2 concentrations. This will speed up the removal of CO_2 from the atmosphere

- Cloud formation. Clouds have a greenhouse effect but also, in part, reflect the visible light. They can hence act as a positive and a negative feedback loop. Cloud formation is stimulated by pollution (especially fine dust particles) and by increased evaporation. Thus, this effect will be more pronounced in industrialising areas

Examples of positive feedback loops include:

- Rising temperatures affect the part of the surface covered by ice, thus reducing the reflection of light. Sea ice and glaciers are retreating in various places such as the Arctic, Kilimanjaro and the Alps[57]

- Rising temperatures stimulate evaporation of ocean water. In the same way as CO_2, water vapour acts as a greenhouse gas

- Rising temperatures may lead to melting of the permafrost (permanent frozen soil) in Arctic regions. This soil consists mainly of organic matter (peat). Under these conditions, oxidation processes could start to create vast amounts of CO_2 or microorganisms could start to generate methane. Recent evidence suggests that the degradation of permafrost is increasing[58]

There are still many uncertainties regarding climate change. Will the ocean absorb (far) more CO_2? If so, what would this mean for sea life? Could we stimulate absorption of CO_2 in the ocean without detrimental effects?

Effects of climate change

- **Rising sea levels** will be a direct threat for people living in low areas. Delta regions, sea marshes and coral reefs are directly endangered; in particular, low-lying island states such as the Maldives[59] and the Marshall Islands could vanish completely. In absolute numbers other regions would be more devastated; 70 million people in coastal China, 70 million people in Bangladesh, 9 million in the Netherlands and about 20 million people in Japan would be threatened. Coastal populations in Indonesia, the Philippines, Malaysia and the USA would also be endangered and ecosystems would suffer. World heritage sites such as the city of Venice would need to enact protective measures. Shore defences would require enormous amounts of capital expenditure, which will probably not be available in poor countries such as Bangladesh. In addition, they might also mean that seasonal river flooding that fertilises the land would no longer be possible

57 NASA, 'Recent warming of Arctic may affect worldwide climate', 23 October 2003; www.gsfc.nasa.gov/topstory/2003/1023esuice.html, 21 October 2005.
58 M.T. Jorgenson, Y.L. Shur and E.R. Pullman, 'Abrupt Increase in Permafrost Degradation in Arctic Alaska', *Geophysical Research Letters* 33 (2006), www.agu.org/pubs/crossref/2006/2005GL024960.shtml, 2 March 2006.
59 www.environment.gov.mv/catas_arti.htm, 21 October 2005.

- **Changing climate zones** will affect agriculture and ecosystems. Rising temperatures and increased droughts will prove devastating for agriculture—especially in areas that already have to deal with water shortages. Droughts might not only reduce the harvest but might also stimulate erosion. Increased CO_2 levels in the atmosphere will stimulate plant growth. The latest IPCC estimates indicate that the situation around the Equator is most likely to worsen, while agricultural production at high latitudes could even increase. The southern rim of the Sahara is becoming greener while the northern rim and southern Europe show signs of desertification[60]

- **Temperatures** keep certain microorganisms from appearing in regions at higher latitudes. These microorganisms might colonise these regions, thus spreading non-indigenous diseases. The situation could worsen as the population might have neither the knowledge nor the cultural or behavioural norms to counteract these diseases. On a global level, the numbers of additional people at risk of malaria in 2080 due to climate change is estimated to be 300 and 150 million for *Plasmodium falciparum* and *Plasmodium vivax* types of malaria[61]

- **Rainfall patterns.** The changes in the climate will also affect the distribution of rainfall during the year. For Western Europe, winter rains are projected to increase while summer rainfall will decrease. This distribution change will increase the risk of winter river flooding

- **Extreme weather conditions** are also likely to increase. The insurance industry is already increasing its premiums due to the effects of the heatwave in Europe in 2003 and the extreme rainfall and flooding in Central Europe in 2002. According to Loster of Munich Re (one of the world's largest reinsurance companies), insurance losses because of weather-related disasters worldwide increased tenfold (adjusted for inflation) between the 1960s and 1990s.[62] Table 2.3 indicates the financial losses due to natural disasters in 2003 and Figure 2.9 shows the increase in such events between 1950 and 2000

- **Melting of glaciers** will occur worldwide if a 4°C rise in temperatures becomes a reality. This will have a profound effect on tourism. Alpine regions have already observed that the lower ski resorts can

60 www.awitness.org/eden_wing/sahara_desert_retreat.html, 21 October 2005.
61 P. Martens, R.S. Kovats, S. Nijhof, P. de Vries, M.T.J. Livermore, D.J. Bradley, J. Cox and A.J. McMichael, 'Climate Change and Future Populations at Risk of Malaria: A Review of Recent Outbreaks', *Global Environmental Change* 9 Supp. 1 (October 1999): 89-107.
62 P. Brown, 'Extreme weather of climate change gives insurers a costly headache', *The Guardian*, 11 December 2003.

	1950–1959	1960–1969	1970–1979	1980–1989	1990–1999	1993–2003
Number of events	20	27	47	63	91	60
Economic losses	42.7	76.7	140.6	217.3	670.4	514.5
Insured losses	–	6.2	13.1	27.4	126.0	83.6

TABLE 2.3 Losses due to natural disasters, 2003 ($ billion)

Source: Topics Geo, *Annual Review Natural Catastrophes 2003*, Münchener Ruckversicherungs Gesellschaft

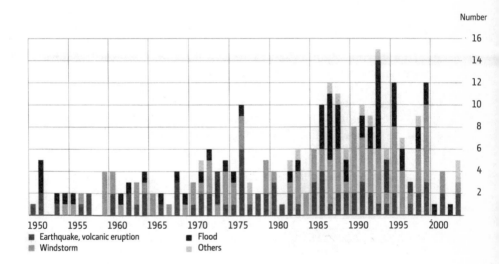

FIGURE 2.9 Numbers of major natural catastrophes, 1950–2000

Source: Topics Geo, *Annual Review Natural Catastrophes 2003*, Münchener Ruckversicherungs Gesellschaft

be used much less. Perhaps more importantly, the melting of glaciers will also affect freshwater supplies. Glaciers act as a storage tank for precipitation, which is released only gradually. Hence they have a stabilising effect on freshwater supplies. Melting glaciers in Arctic regions (especially Greenland) may influence the local salinity and density of the ocean and thus affect the Gulf Stream. This might have a considerable climatic effect on North and Western Europe[63]

Ozone layer

Solar radiation contains various harmful forms of radiation—particularly ultraviolet-B (UV-B), with wavelengths of 280–320×10^{-9} m (frequency 1×10^{15} Hz). A large part of UV-B is destroyed by the ozone layer. A thinner ozone layer means that more UV-B reaches the surface of the Earth. The intensity of UV-B depends on the thickness of the ozone layer, latitude, time of the day and season.[64]

DNA absorbs UV-B and can thereby be damaged. For humans the effects can include skin irritation, less synthesis of vitamin D, degradation of the immune system and skin cancer. Increased UV-B levels also affect animals and plants; phytoplankton in Arctic areas are particularly vulnerable. The consequences of decreased phytoplankton growth might be severe as it is the basis for an important food chain.[65] Studies on the effect of increased UV-B levels for marine wildlife show that there is reason to be worried. Agriculture can also be affected by increasing UV-B; harvests will be lower—though the extent is hard to predict as various factors coincide. Figure 2.10 shows the concentrations of ozone in the atmosphere.

At the beginning of the 1970s, Rowland and Molina[66] studied the degradation of chlorofluorocarbons (CFCs) in the atmosphere. They drew an alarming conclusion:

> CFCs in the atmosphere are only destroyed by UV radiation. High doses of UV only occur in the stratosphere, in the ozone layer. The destruction of CFCs produces chlorine ions that act as catalyzing agents for the destruction of ozone molecules. As a result, more UV-B radiation will reach the surface of the earth.

By 1974, CFCs had a long history of use in industry. They were developed as refrigerants for refrigerators and air-conditioners. Later they were used as propellants for spray cans. Because CFCs are completely inert, they turned

---

63 'Melting glaciers diminished Gulf Stream, cooled Western Europe during last Ice Age', National Science Foundation (NSF) press release, 19 November 2001; www.nsf.gov/od/lpa/news/press/01/pr0192.htm, 21 October 2005.

64 Ozone is present at high altitudes (the stratosphere) and low altitudes (the troposphere). Tropospheric ozone causes lung problems, while stratospheric ozone protects us from UV-B.

65 C. Woodward, 'Food-chain Alarm from a Low-ozone Zone', *Christian Science Monitor*, 11 December 1998; www.csmonitor.com/durable/1998/12/11/p8s1.htm, 21 October 2005.

66 M.J. Molina and F.S. Rowland, 'Stratospheric Sink for Chlorofluoromethanes: Chlorine Atom Catalysed Destruction of Ozone', *Nature* 249 (1974): 810-12.

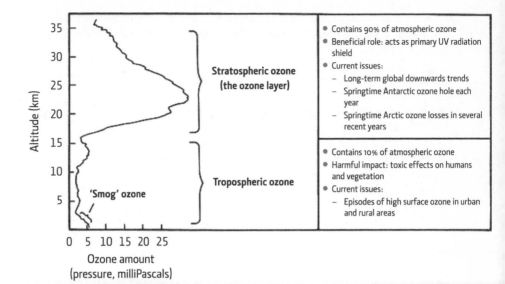

FIGURE 2.10 Distribution of ozone in the atmosphere

Source: www.ndsc.ncep.noaa.gov/freq_qu/faq.html, 25 October 2005

out to be ideal as blowing agents to make plastic foams, as gaseous insulator for transformers, and as a de-fluxing agent in the electronics industry (to clean printed circuit boards).

Although depletion of the ozone layer in 1976 was a phenomenon that only showed up in mathematical models and lacked any empirical evidence, public concern was high and sales of spray cans slumped. However, the sense of urgency regarding the ozone issue declined and, with the election of Ronald Reagan as US president in 1980, efforts to regulate the use of CFCs ceased.

This situation changed completely with the publication of an article in *Nature* on 16 May 1985.[67] This reported on very strong thinning of the ozone layer measured above the Antarctic between August and November 1984. The public concern that was raised allowed UNEP to make a leap forward on the ozone issue. The Montreal Protocol on Substances that Deplete the Ozone Layer was signed on 16 September 1987 by 27 countries; it has subsequently been amended several times. The Protocol encompassed:

- A total ban of CFCs by the end of the 20th century

---

67   J.C. Farman, B.G. Gardiner and J.D. Shanklin, 'Large Losses of Total Ozone in Antarctica Reveal Seasonal ClO_x/NO_x Interaction', *Nature* 315 (1985): 207-10.

- A fund for developing nations to help them to acquire and introduce new technologies that can replace those that use ozone-depleting substances
- A ten-year grace period for developing nations to comply with the protocol

Although many ozone-depleting substances are still on their way up to the ozone layer, it is assumed that the ozone layer will recover in about 50 years. The ozone problem therefore shows that it is not impossible to achieve results by international co-operation.

Air pollution

Acid rain is one of the worst pollution problems. Its most important source is the burning of fossil fuels. The burning of fuels with a high sulphur content has resulted in the destruction of forests—especially in the mountains of Central Europe.

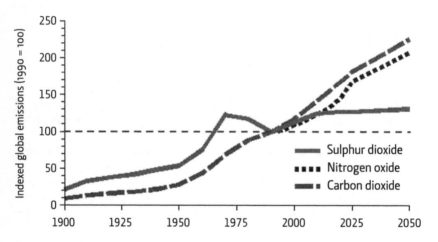

FIGURE 2.11 **Emissions of the three main air pollutants**

Source: UNEP, *Global State of the Environment Report 1997* (GEO-1), 1997; www.unep.org/Geo/geo1/fig/fig4_6.htm, 21 October 2005

The burning of fossil fuels can also affect human health. One famous example is the London smog, which in December 1952 killed an estimated 12,000 people and brought illness to many others.[68]

68 www.portfolio.mvm.ed.ac.uk/studentwebs/session4/27/greatsmog52.htm, 21 October 2005.

According to the US Environmental Protection Agency, elderly people especially die prematurely each year from exposure to ambient levels of fine particles.[69] Also, because children breathe 50% more air per pound of body weight compared with adults, they are more susceptible to tiny particles in the pollution—especially if they suffer from asthma.[70]

Noise

Noise is a problem that many inhabitants of densely populated areas have to live with. Thin walls, urban traffic or the vicinity of motorways, railways or airports can lead to high levels of noise. Living amid high noise levels has often no direct effect but is a stress factor in the longer term as it can lead to less sleep. High noise levels can also create hearing problems. Noise is often an important issue—especially around airports.

Lithosphere

The lithosphere is the crust of the Earth. It provides us with various substances and energy through mining and it acts as the basis for the biosphere.

Mining

The crust of the Earth contains various materials:

- Metals such as iron, aluminium, zinc, copper, gold, silver and various rare earths

- Salts such as sodium chloride

- Fossil fuels such as methane, coal and crude oil

- Sedimentary rock for construction

- Minerals such as asbestos, calcite ($CaCO_3$), dolomite ($CaMg(CO_3)_2$), fluorspar, feldspar and phosphate

- Elements such as beryllium, bromine, diamond (a form of carbon) and silicon

The temperature of the Earth increases with depth. This phenomenon is sometimes also used for energy production, particularly for warming houses as in Iceland (geothermal energy).

The lithosphere contains 99.99% of all carbon: there is far more carbon in the lithosphere (in the form of mineral rock) than in fossil fuels (see Table 2.4).

---

69 US EPA, 'Particulate Matter: Health and Welfare', www.epa.gov/air/particlepollution/health.html, 17 March 2006.
70 www.eurekalert.org/releases/udel-siho81099.html, 13 April 2005.

Sink	Amount (billions of tonnes)
Sediments and rocks	66,000,000–100,000,000
Ocean	38,000–40,000
Biosphere—soil	2,000–2,200
Fossil fuels	4,000
Atmosphere	766 (in 1999)

TABLE 2.4 Estimated major stores of carbon on the Earth

Source: M.E. Ritter, 'The Physical Environment: An Introduction to Physical Geography';
www.physicalgeography.net/home.html, 21 October 2005

Not all fossil fuels can be explored. Some deposits are too deep or in hard-to-reach places. Other deposits are too small or otherwise neither technically nor economically feasible for exploration. Of the deposits that are actually explored, not all fuels can be recovered. For example, as much as 98% of the oil deposit might not be reached if only mechanical pumps were used to recover oil. Enhanced oil recovery techniques allow 20–60% of the content of an oil deposit to be recovered.[71]

The term 'economically recoverable reserves' is generally used when giving estimates of the available deposits of fossil fuels, minerals and ores. These are the reserves that can be recovered economically at current prices and using current technology. With increasing prices and improved recovery technology, these reserves may grow much bigger without any new deposits being found!

How long will our fossil fuels last? There is no final answer to this question. Oil could be produced for another 40 years at current rates of production, at current oil prices and using current technology.[72] However oil consumption has risen about 13% in the last decade, new reserves will be found and technology might be improved. This will influence the market and the resulting market prices will influence consumption, exploration of new oil fields and investments in technology. Still we must be prepared for a shortage of oil and one that could even happen within 50 years. For natural gas, the proven reserve is 60 years, although in this case, consumption is rising much faster, i.e. about 20% in the last decade. Proven reserves of coal will last for another 204 years; coal consumption grew 8% in the last ten years.[73]

71 N. Bunce and J. Hunt, 'Enhanced Oil Recovery'; www.physics.uoguelph.ca/summer scor/articles/scor134.htm, 21 October 2005.
72 See BP Statistical Review of World Energy 2005; www.bp.com/genericsection.do?categoryId=92&contentId=7005893, 21 October 2005.
73 BP Statistical Review of World Energy 2005; www.bp.com/genericsection.do?categoryId=92&contentId=7005893, 21 October 2005.

The same story holds for ore mining. As prices rise or technology improves, less-rich ores can be mined and thus the reserves increase. However, various metals can be recycled—especially when prices rise and recycling technology improves. Metal ore mining is therefore likely to decrease.

Mining can have various hazardous effects:

- Shafts can collapse

- Mines can catch fire. Sometimes these fires cannot be extinguished

- Water can seep through mines. The resulting acid drainage water can result in streams that are entirely devoid of life

- Open-pit mines can wipe away the entire landscape and leave barren pits thousands of feet deep. The drainage of these pits can lead to desertification for considerable distances

- The debris from mines can clog the landscape. Precipitation falling on this debris can become acidified and/or toxic

Topsoil layer

There are also a number of serious problems at the Earth's surface. Industrial activities, which have sometimes taken place many centuries ago, have polluted the soil and the groundwater at various sites around the world. For example, town gas works are notorious for polluting soil. Chemical and even nuclear waste sites can create great dangers, especially in geologically unstable regions of the world.[74] In industrial countries, industrial practices have improved considerably, but the contamination from the past is being cleared very slowly as the costs are enormous.

Municipal waste is still often landfilled. Besides the land that landfills occupy, they may pollute the precipitation that washes through them, create odour and produce considerable amounts of methane (a greenhouse gas). This should be flared off or, even better, used in productive processes. Landfill leachate and percolate should be collected and treated.

Agriculture can also damage the soil. The excessive levels of minerals in cattle manure are gradually poisoning the soil. Especially meadow land that is intensively used in dairy production often has too high a content of heavy metals and phosphates due to fertiliser use and the external supply of food to the cattle (which ends up on the land in the dung). Throughout Europe,

74 Cf. I. Hadjambardiev, 'A Sustainable Development Course for Environmental Engineers in Kyrgyzstan', *International Journal of Sustainability in Higher Education* 5.3 (2004): 289-94.

highly productive agricultural land will become unproductive within a century due to toxicity through the increased the levels of these minerals.[75]

In various places, agricultural activities cause erosion. Loss of the fertile topsoil can create a permanent desert and lead to food shortages.

Questions, discussion and exercises

1. Search the internet for information about the last weather-related disaster and estimate the damages. How many large-scale weather-related disasters can you find for the last two years?

2. Estimate the amount of additional CO_2 in the atmosphere compared with a century ago. What is the equivalent amount of hydrocarbons (C_nH_{2n})?

3. What are the priorities in the national environmental policy in your country/region? What were they ten years ago? Pick a country on a different continent and search the internet for information on its priorities. Try to explain the differences.

4. Estimate the rise in the average sea level based on the assumption that the ocean is 5 km deep, the water temperature is 6°C and the water temperature increases by 1°C.

75 M. Boer and K-J. Hin, 'Zware metalen in de melkveehouderij: Resultaten en aanbevelingen vanuit het project Koeien & Kansen'. *Koeien & Kansen Rapport* 16, CLM-nr.587-2003, 2003; www.koeienenkansen.wageningen-ur.nl/index.asp?media/rapport/16.asp, 21 October 2005.

3 Patterns of development

Sustainable development does not imply just dealing with the physical problems that threaten the continuity of civilisation on our planet. It also implies development. But, in thinking of development, it is easy to fall into the pitfall of linear thinking, i.e. supposing that every nation passes similar stages leading to societies such as those in North America, Europe and Japan.

This chapter outlines, in a historic fashion, the development processes of various societies. It focuses on the role that technology has played in regard to the organisation of these societies and its exchange with the life-sustaining system.

A brief history of global development

The development of human societies can be classified in four stages:

- Hunters and gatherers
- Agriculture
- Industrial production
- Global patterns of production and consumption

These four stages are considered below.

Societies neither pass through these stages with equal speed nor are the stages strictly linear. Barriers and leaps sometimes occur. Development at various remote places increasingly influences that at other places.

First stage: hunters and gatherers

Humans are relatively new inhabitants of this planet, appearing about two million years ago when biodiversity had reached a peak. These people lived as nomads in groups of probably 30–40.[1] They obtained about two-thirds of their food from the fruits and vegetables they gathered and one-third from the meat of wild animals. Their only energy source was fire. Dogs were probably the only domesticated animals.

This nomadic way of living required considerable space from which to collect food. The hunter–gatherers gradually spread all around the planet and occupied most of the favourable environments. The population density was very low and depended on the productivity of the local ecosystem. The Earth was therefore only able to sustain between 10 and 30 million people. On average, people did not have to work hard for their living—a couple of hours per day were probably enough. They had plenty of time to spend on exercise or cultural activities. Infectious diseases were absent as human contact was too sparse to enable these to spread.

However, life was not as good as this might sound. Life was very dangerous for these nomads. Many of them died of injuries inflicted on them by wild animals; alternatively, they suffered from these injuries for the rest of their lives. There was not always cover to protect them from heat, cold or precipitation. Occasionally, hunger or drought took their toll.

This way of life was the dominant mode for about 50,000 generations and can still be found in some parts of the world such as the Amazon forests and Irian Jaya (the western half of Papua New Guinea, Indonesia).[2]

This way of life is environmentally sustainable as human behaviour is completely integrated into the ecosystems. But the capacity of the Earth's ecosystems allowed no more than about 30 million people to live in this way.

Second stage: agriculture

When the population of hunters and gatherers rose above the regional 'eco-capacity' and all habitats had been colonised, competition for food began. Nomad groups began to settle and, naturally, claimed the most productive locations. People were therefore driven to begin enhancing the productivity of ecosystems.

Humans domesticated nature to enhance its productivity. This became known as the agricultural revolution and encompassed the domestication of plants and cattle. This reduced direct dependence on nature as people could now plan their production (more or less), but it increased the amount of work.

Productivity could be further increased by:

1 R.E. Leakey and R. Lewin, *Origins Reconsidered* (New York: Doubleday, 1992).
2 www.irja.org/index2.shtml, 2 March 2006.

- Clearing forests
- Constructing various structures such as water basins and ditches to collect water in areas with only seasonal rainfall and distribute it in dry seasons
- Constructing drainage systems to reclaim fertile marshlands

The first impacts on nature followed. Landscape transformation on a local scale included:

- Local erosion
- Salinisation of soil due to irrigation
- Local eradication of animals that affected agricultural production

Moreover, the way of life became sedentary, i.e. people remained at the same place for the major part of their lives. This led to specialisation of labour as people could increase production according to their skills or local available resources. However, labour specialisation and trade can lead to social inequality.

Non-agricultural production was no longer tied to the land. Cities could arise as trading centres and as political and military centres of power. To maintain the collective infrastructure (e.g. city walls and gates, military, water infrastructures), taxes were introduced and administrative systems were created. This was only possible following the invention of writing.

Cities also became cultural centres, with theatres and street plays. But they depended on the surrounding countryside, which they dominated, for food supply. Longer-distance trade was rare and confined to high-value products.

New technologies developed (e.g. ceramics, writing and metalworking). In these new societies, new phenomena appeared such as:

- Social inequalities
- Larger-scale wars
- Epidemics
- Goods and services markets
- Commercial routes
- Slavery
- Democracy
- Sciences
- Institutionalised religions

This stage lasted for some 400 generations.

Third stage: industrial production

Increased productivity allowed for greater population growth. But, because the population was sedentary, this growth was unequal. Cities were fed by their surrounding countryside but, as production in cities also required a supply of raw materials, trade increased.

Trade routes, which thus far had been used only for valuables, were increasingly used for non-perishable goods such as wood and wheat. Unknown territories were explored and trade routes were set—the most important being the sea route between Europe and Asia.

Local overpopulation also led to emigration. In regions where the indigenous population was still hunting and gathering such as in North America, immigrants were able to easily replace or enslave this population and set up their own communities. Contagious diseases introduced by the Europeans decimated the agricultural societies of Central America and the Andes, and the remaining people were suppressed by force.

The use of fossil fuels further improved productivity in cities. The invention of the Newcomen machine in 1711 (Figure 3.1) and the steam engine in

FIGURE 3.1 **Newcomen machine,**
1711

Source: J. Leupold, Theatrum machinarum Hydraulicarum Leipzig, 1725

1762 allowed fossil fuels to be used not only for heating but also for the mechanisation of production.

The steam engine had other advantages. It increased the availability of fossil fuels considerably as mines could be drained far more effectively. Moreover, the fuels themselves, raw materials and commodities could be transported more quickly using steam-powered trains and boats. Increased availability of fuels and better transport facilities enabled cheap large-scale industrial production of commodities based on the supply of raw materials from all around the world.

The trade relations developed by European countries in the 16th–18th centuries with more advanced agricultural societies such as India, Indonesia and China gradually changed. The economies in these regions were re-ordered to serve the resource needs of the colonial powers. This colonial trade not only supplied the raw materials for industrial production, but the trade in these materials also supplied the capital needed for this expansion.

Later, major European countries secured these supplies by taking over countries that had initially been their trading partners. The main colonial powers were:

- UK (e.g. Indian subcontinent, East Africa, Nigeria and Australia)
- Netherlands (e.g. Indonesia)
- France (e.g. North Africa and Indochina)
- Portugal (e.g. Angola, Mozambique and Brazil)

Smaller amounts of territory were held by Belgium, Germany and Spain (after it lost South America in the early 19th century). Russia also expanded in a colonial manner into Central Asia and Siberia.

Through the emergence of the colonial system, trade had evolved into exploitation. The forced shipment of about 15 million African slaves to the Americas between 1540 and 1850 was one of the darkest aspects of world history. Large amounts were earned through this 'trade' and the cheap labour allowed the USA to expand. The USA (a former colony of the UK) conquered the Philippines in 1898. Japan (Korea) also set up colonies. Colonialism peaked before the outbreak of World War II in 1939 (Table 3.1).

Industrial production began in the UK during the 18th century and was soon followed by various countries in continental Europe (Belgium, Germany, Austria and France). By the end of the 19th century, the USA, Japan and other European nations had begun to industrialise.

The textile sector was the first to adopt industrial methods. Even in the colonies, large steam-powered spinning and weaving mills replaced small-scale local production. Steelworks, coal mines, engineering and shipbuilding were the next emerging sectors of industrial production. In addition to coal, oil and gas were exploited as fossil fuels as they were easier to obtain and transport.

The main driving forces of this third stage were:

	Great Britain	France	Belgium	Netherlands	Germany (1914)
Area (square miles)	94,000	212,600	11,800	13,200	210,000
Population	45,500,100	42,000,000	8,300,000	8.500,000	67,500,000
Area of colonies	13,100,000	4,300,000	940,000	790,000	1,100,000
Population of colonies	470,000,000	65,000,000	13,000,000	66,000,000	13,000,000

TABLE 3.1 The extent of colonialism in 1939

Source: M.E. Townsend, *European Colonial Expansion since 1871* (Chicago: J.P. Lippincott Company, 1941): 19; www.mtholyoke.edu/acad/intrel/pol116/colonies.htm, 8 November, 2005

- The introduction of fossil fuels as a source of energy
- The growth of the railway system
- The growth of overseas trade

At the beginning of the 20th century, new branches of industry were added (electrical equipment, chemistry and car manufacturing). This stage lasted only eight generations.

Fourth stage: global patterns of production and consumption

At the beginning of the 20th century, research and development (R&D) became increasingly important for various branches of industry.

The dyeing industry was the first sector to recognise the importance of science. It allowed them to produce chemical dyes that were cheaper or brought hitherto unavailable colours to textiles. Bayer AG in Leverkusen, Germany, was the first company to set up a chemical laboratory, but was quickly followed by its German competitors. This move contributed to Germany's world-class status in chemistry in the early 1900s.

The emerging electrical industry was itself a product of research. The company set up by Thomas A. Edison, General Electric, was the first electric company to set up a research department.[3] It was followed by Westinghouse (an early competitor) and sometime later by Siemens and Philips in Europe. Du Pont was successful in gunpowder research and Courtaulds in fibre research.

Businesses that became dependent on R&D tended to merge in order to create larger companies. This was because R&D needs to be done at a certain minimal scale, which small companies tend to find difficult to achieve. Those industries that depended on R&D each developed in their own terri-

3 T.P. Hughes, *Networks of Power: Electrification in Western Society, 1880–1930* (Baltimore, MD: Johns Hopkins University Press, 1983).

tory. A good example of these divided markets can be seen in the chemical industry. Before World War II, the US company E.I. Du Pont de Nemours dominated various parts of chemical business in the USA (e.g. fibres, explosives and refrigerants such as chlorofluorocarbons [CFCs]). Latin America was its natural area of expansion. Du Pont succeeded in dominating the chemical market there and set up several production facilities.[4] As is clear from the name, Imperial Chemical Industries had the same role in the British Empire. In large parts of continental Europe, IG Farben (formed through the merger of the largest German chemicals producers) dominated major parts of the chemical business in Central and Eastern Europe. As these large multinational companies rarely competed with each other, they could reduce costs by swapping the new products they developed through their R&D.[5]

Two developments terminated this co-operation:

- Governments took decisive action against monopolies and trusts (especially in the USA). Attractive and protected markets were therefore no longer guaranteed

- Internationally, the USA was successful in destroying trade barriers[6]

The result was that companies were forced to compete at home and started to penetrate each other's markets. At this point, they became multinational.

The importance of economies of scale in major industrial sectors was a main driver for globalisation. In the 1970s, companies discovered that it paid to concentrate on specific technology areas and markets. In these areas, a company could be a large player with considerable market power. Thus, large industrial conglomerates gradually became larger and smaller at the same time. With this strategy, they were best able to obtain a high return on the fixed costs of product development and improvement. The resulting multinationals are neither larger nor smaller than they were in the 1960s, but they are more focused on specific technologies and markets.

Globalisation does not mean more global trade. Global trade is not growing faster than the average growth of the world economy. In fact, trade volumes were relatively higher during the colonial times when manufactured goods were produced by the industry of the colonial power and the raw materials were generally produced in the colonies.

---

4   D. Hounshell and J.K. Smith, *Science and Corporate Strategy: Du Pont R&D 1902–1980* (Cambridge, MA: Cambridge University Press, 1988)
5   F. Aftalion and O.T. Benfey, *A History of the International Chemical Industry* (Philadelphia, PA: University of Pennsylvania Press, 1991).
6   The political driving force was probably a combination of free-market ideology and anti-communism, and US national interest (as the strongest industrialised nation).

Globalisation also does not mean global flows of capital. Large companies prefer to loan the capital that they need in the countries where they need it.[7] Globalisation means that the more or less fixed costs of a technological design are earned back by selling worldwide. In particular, products with low transport costs can easily be marketed worldwide. Products with higher transport costs can be produced in the main regions of the world.

Companies that are unable to operate worldwide have the choice of finding overseas partners or merging. The result can be observed by every traveller. Some decades ago, it was easy to buy interesting 'local' presents for people back home. Nowadays, it is much harder as the 'gifts' on offer appear identical in every souvenir shop.

Intermezzo: from autarchy[8] to globalisation—the example of Ibieca

Ibieca is a rural village situated 23 km from the city of Huesca in north-eastern Spain. In 1996, there were 110 Ibiecanos and 183 temporarily registered inhabitants (i.e. city-dwellers who lived there for the weekend).

In 1900, 90 families lived in Ibieca. The main element of social order was the 'casa', which consisted of the family house, its members and its livestock. Ibieca was more a community of casas than of individuals.

Three generations often lived in one house. The prosperity of the casa (the 'measure' of success) was reflected in the size of the house; most were two-storey houses, with three-storey houses for the wealthier families and one four-storey house for the Solana family, which owned nearly 300 hectares.

There was considerable inequity in the village. Those from casas with no land and those belonging to small casas offered themselves as paid labour to the larger casas. Medium-sized casas sometimes hired labour and sometimes offered labour depending on the season and the availability of labour within the casa. Although this inequity produced conflict,[9] there were strong bonds of loyalty towards the 'amos'—the rich landowner/employer.

Casas could extend their income by entrepreneurial activities such as trade or transport services. The village as a whole was to a large extent autarchic. It produced:

---

7   J. Wengel and A. Kleinknecht, 'The Myth of Economic Globalisation', *Cambridge Journal of Economics* 22.5 (1998): 637-47.
8   'Autarchy' means self-sufficiency of a region or country.
9   During in the Spanish Civil War (1936–1939), there was about a 50:50 split between those in favour and those who opposed the republicans (various leftist groups that supported the Spanish republic, in the Ibieca region mainly anarchists) who controlled the region and their reorganisation of agriculture.

- Cereals that were:
  - Ground to produce flour
  - Baked to make bread
  - Used as feed for the animals
  - Sold
- Grapes from which wine was produced
- Olives that were ground to produce cooking oil and lamp oil
- Fruits and vegetables
- Almonds
- Flowers
- Cane used for baskets and chairs
- Hemp
- Firewood, charcoal and acorns
- Soil for construction materials
- Water
- Donkeys, mules and oxen for transport and ploughing
- Sheep for wool and meat
- Goats for milk, cheese and their skins
- Rabbits, chickens and pigs for meat
- Pigeons and chickens for meat and eggs
- Cats and dogs

Only almonds were produced solely for the market, though surpluses of cereals, skins, wine, olive oil and wool were also marketed. Village households consumed most of what they produced and produced most of what they consumed. Those produced items that were not consumed were often traded regionally, e.g. with villages in the Pyrenees that could not produce wine. This trade was one of the activities through which villagers could extend their income.

There was almost no material waste. Every product was harvested and even the spilled grains were recovered. Every part of a slaughtered pig was made into food. Cats survived on rats and mice, and dogs and donkeys received scraps.

The land was divided into many plots. Although sometimes uneconomical, this practice had the advantage that not all of a crop was destroyed by severe weather conditions (frost, hail, etc.).

After Franco won the Civil War, it was important to secure food supplies. The trade ban instigated by the World War II allied powers led to a food shortage. State regulation of food production and a rise in wheat prices stimulated the transition of land use towards wheat production.

In 1950, the USA acknowledged Spain as an ally in the Cold War. Trade bans were lifted and international loans were made available. The Spanish Government made loans and subsidies available to farmers to purchase machinery and modern buildings. Reservoirs were built and the reallocation of farmland to create larger plots was encouraged.

In Ibieca, there were four tractors in 1958, seven tractors and one combine harvester in 1967, and 12 tractors and five combine harvesters in 1975. The transition to monoculture and the mechanisation of work made Ibieca part of the world economy: Many people left Ibieca as this had also created unemployment, and falling prices led to the bankruptcy of marginal casas.

Practically everything that is produced in Ibieca today is not consumed there and the people of Ibieca buy everything they need (not in Ibieca, but in Huesca). The houses belonging to the Ibiecanos who left the town are now occupied by city dwellers as weekend retreats.[10]

Key issues in the development of society

What is development?

Are the four phases described above development? Anti-globalisation demonstrators would not agree and many others would have second thoughts.

Speaking about development implies that you have a goal to work towards. Development can be measured only by measuring progress. Without objectives, we might just speak of 'change'.

The objectives that humanity pursues are very different in the various cultural traditions, religions and ideologies that act as guiding principles for people or societies. However, there are some principles that most of us share. On 10 December 1948, the General Assembly of the United Nations adopted and proclaimed the Universal Declaration of Human Rights.

Are there core elements in development that are more or less universally valid? In his book published in 1971, Denis Goulet discerned three basic components or core values:[11]

> *Life sustenance: the ability to provide basic needs.* All people have basic needs which have to be met otherwise life becomes impossible: food, shelter, health, protection (from violence, or threat of

---

10 S.F. Harding, *Remaking Ibieca: Rural Life in Aragon under Franco* (Chapel Hill, NC: University of North Carolina Press, 1984).
11 D. Goulet, *The Cruel Choice: A New Concept in the Theory of Development* (New York: Atheneum, 1971).

Preamble to the Universal Declaration of Human Rights[12]

- Whereas recognition of the inherent dignity and of the equal and inalienable rights of all members of the human family is the foundation of freedom, justice and peace in the world,

- Whereas disregard and contempt for human rights have resulted in barbarous acts, which have outraged the conscience of mankind, and the advent of a world, in which human beings shall enjoy freedom of speech and belief and freedom from fear and want has been proclaimed as the highest aspiration of the common people,

- Whereas it is essential, if man is not to be compelled to have recourse, as a last resort, to rebellion against tyranny and oppression, that human rights should be protected by the rule of law,

- Whereas it is essential to promote the development of friendly relations between nations,

- Whereas the peoples of the United Nations have in the Charter reaffirmed their faith in fundamental human rights, in the dignity and worth of the human person and in the equal rights of men and women and have determined to promote social progress and better standards of life in larger freedom,

- Whereas Member States have pledged themselves to achieve, in co-operation with the United Nations, the promotion of universal respect for and observance of human rights and fundamental freedoms,

- Whereas a common understanding of these rights and freedoms is of the greatest importance for the full realisation of this pledge.

violence). If these needs cannot be met, a condition of absolute underdevelopment exists.

Self-esteem: to be a person. All people seek some form of self-esteem, which they might call identity, dignity or authenticity. As long as esteem or respect was dispensed on grounds other than material achievement, it was possible to resign oneself to poverty without feeling disdained [. . .] Nowadays the Third World seeks development in order to gain the esteem which is denied to societies living in a state of disgraceful 'underdevelopment'. Today, we could add that many countries have not only a low self-esteem because of underdevelopment,

12 Universal Declaration of Human Rights. Adopted and proclaimed by UN General Assembly Resolution 217 A(III) of 10 December 1948; www.un.org/Overview/rights.html, 8 November 2005.

but also because of the outright disdain by which their culture is portrayed in the rich world.

Freedom from servitude: to be able to choose. This is not to be narrowed down to political or ideological freedom. It means that persons and societies expand the range of choices that they are able to make, and minimise the external constraints. Economic growth is not primarily important for creating wealth, but for creating the possibility for choice.

The UN has developed further principles regarding goals for development. One important element for development, which many regard as the only one, is economic production. This is measured by gross domestic product (GDP), which is generally expressed in US$ or US$ per capita. However, GDP is often a poor measure of wealth because:

- GDP counts only financial transactions. Production that is no part of the economy (e.g. the people of Ibieca producing mainly for themselves) is not included

- Currency exchange rates are often unrealistic. The purchasing power of one dollar is far greater in India than in New York. An interesting attempt to compensate for this is made by *The Economist*'s Big Mac index[13] (see example in Figure 3.2), in which currency exchange rates are recalculated based on the number of hamburgers that could be purchased using the local currency. However, hamburgers are often only for sale in the capital city of developing countries (which tends to be richer) and are consumed only by a small wealthy minority

GDP is a measure for production and not income. Various poor countries are highly indebted. A considerable part of their income is used to pay interest on loans (see below).

Moreover, should material production be the ultimate goal of development? Other goals could be:

- The participation of the whole population in the main processes of society

- Absence of racial or sexual discrimination

- Good conditions of living for everyone

This range of goals for development can be seen in the eight UN Millennium Development Goals:[14]

13 www.economist.com/markets/Bigmac/Index.cfm, 8 November 2005.
14 www.un.org/millenniumgoals, 8 November 2005.

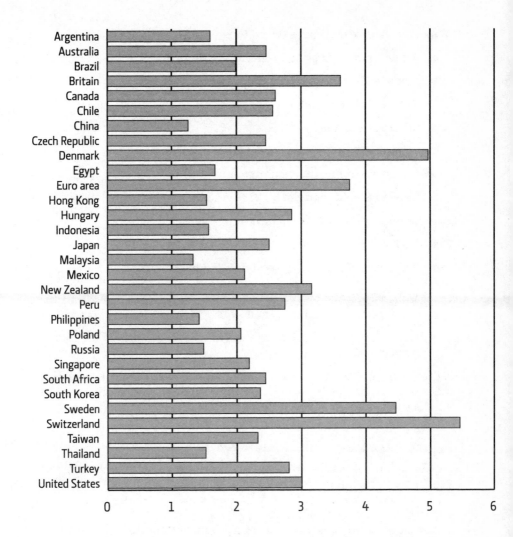

Figure 3.2 Big Mac index, 13 December 2004

Source: *The Economist*, 16 December 2004; based on McDonald's price data. Big Mac prices in US$ per country;
www.economist.com/markets/bigmac/displayStory.cfm?story_id=3503641, 2 March 2006

- Eradicate extreme poverty and hunger
- Achieve universal primary education
- Promote gender equality and empower women
- Reduce child mortality
- Improve maternal health
- Combat HIV/AIDS, malaria and other diseases
- Ensure environmental sustainability
- Develop a global partnership for development[15]

To measure progress, you could focus on:

- Literacy
- Participation in primary/secondary/higher education
- Unemployment
- Life expectancy (resulting from healthcare and absence of famine)
- Percentages of males/females and minorities in various key functions in society

Each year, the UN Development Programme (UNDP) calculates various indexes based on these parameters. These indexes are published in its annual Human Development Reports.[16]

Population growth

Population growth has been an important driving factor in human history. Exponential population growth began with the industrial revolution. However, the fastest growth took place in the second half of the 20th century, when the population doubled.

Population growth is not distributed equally; the highest growth is achieved in the countries in the southern hemisphere and urban areas have the highest growth rates. Half the global population was living in urban areas at the end of the 20th century. The average urban inhabitant uses far more resources than someone living in the countryside. This is mainly due to the 'agricultural stage' lifestyle still present in many rural areas.

There is a strong relation between birth rates and wealth. In 1992, Meadows and colleagues[17] concluded that countries with higher GDPs have lower

15 www.developmentgoals.org, 8 November 2005.
16 hdr.undp.org/reports, 8 November 2005.
17 D.H. Meadows, D.L. Meadows and J. Randers, *Beyond the Limits: Confronting Global Collapse, Envisioning a Sustainable Future* (White River Junction, VT: Chelsea Green Publishing Company, 1992).

birth rates while countries with lower GDPs have higher birth rates. Poverty and population growth are related: poverty brings population growth and the population growth brings more poverty.

Governments of under-developed nations face a difficult problem as the rationality of high reproduction rates at the micro level is often clear. Children are cheap labour that can help the family to survive. Later on they can take care of their parents. As infant mortality is high, reproduction should even be higher. When infant mortality drops (thanks to better healthcare), reproduction rates gradually follow suit.

However, it is not that easy. Writing more than 200 years ago in his *Essay on the Principle of Population*,[18] Thomas Malthus described what he called the 'population trap'—the economic growth that is achieved might (with a time delay) lead to an even higher population growth. The net economic growth that is achieved will, in this way, produce a net decline of per capita income, i.e. the country is trapped in its poverty. Birth control is therefore important for economic development.

Global population growth rate has fallen in recent years. A major factor in this decline is AIDS, which has eliminated the previously expected population growth in sub-Saharan Africa almost completely and affected it considerably in other countries.[19] The slower pace of population growth will not improve the opportunities for long-term development in these regions. The most productive generations are generally hit most—causing not just a human tragedy, but an economic disaster as well.

Birth rates are going down in most industrialised nations. Some European countries have declining populations, but immigration means that the population of the USA and Canada is still rising.[20] Table 3.2 and Figure 3.3 show how the world's population is expected to grow up to 2050.

Urbanisation

Throughout history the distribution of the population has become increasingly concentrated. As autarchic production became less important, an increasing part of the population was employed in the manufacture of commodities. These people were concentrated in the cities.

Throughout history cities were also political centres that provided protection for their inhabitants. People living in cities (especially those in capitals) are still more important political allies for any government than those less educated people living far away from the centres of power.

18 www.ac.wwu.edu/~stephan/malthus/malthus.o.html, 8 November 2005.
19 United Nations Department of Economic and Social Affairs, *Population, Development and HIV/AIDS with Particular Emphasis on Poverty: The Concise Report* (New York: United Nations, 2005; available at: www.un.org/esa/population/publications/concise2005/PopdevHIVAIDS.pdf, 2 March 2006).
20 F. Pearce, 'Global population forecast falls', *New Scientist*, 27 February 2003; www.newscientist.com/news/news.jsp?id=ns99993444, 8 November 2005.

	Africa	Asia	Europe	Latin America	North America	Oceania	WORLD
2005	887,964	3,917,508	724,722	558,281	332,156	32,998	6,453,629
2010	984,225	4,148,948	719,714	594,436	348,139	34,821	6,830,283
2015	1,084,540	4,370,522	713,402	628,260	363,953	36,569	7,197,246
2020	1,187,584	4,570,131	705,410	659,248	379,589	38,275	7,540,237
2025	1,292,085	4,742,232	696,036	686,857	394,312	39,933	7,851,455
2030	1,398,004	4,886,647	685,440	711,058	407,532	41,468	8,130,149
2035	1,504,179	5,006,700	673,638	731,591	419,273	42,803	8,378,184
2040	1,608,329	5,103,021	660,645	747,953	429,706	43,938	8,593,592
2045	1,708,407	5,175,311	646,630	759,955	439,163	44,929	8,774,395
2050	1,803,298	5,222,058	631,938	767,685	447,931	45,815	8,918,725

TABLE 3.2 Expected world population to 2050

Source: United Nations Secretariat, Division of the Department of Economic and Social Affairs, *World Prospects: The 2002 Revision and World Urbanisation Prospects: The 2001 Revision*; esa.un.org/unpp, 8 February 2005

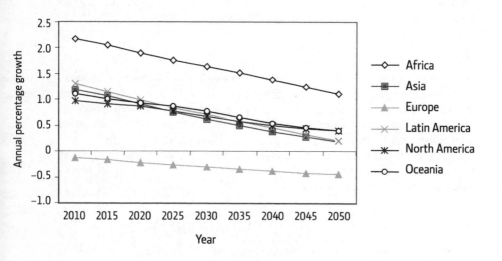

FIGURE 3.3 Population growth rates (annual percentage)

Thus, metropolitan citizens often enjoyed greater privileges than peasants living in countryside; in many cases, food prices mean that this is still the case. Low food prices can keep the city population at peace with the ruling party and many governments regulate food prices. As a result, the farmers producing the food for the city find it difficult to earn a decent living. The city with its low food prices and potential for higher income is a powerful attraction for migration. When this situation evolves into famine, foreign food aid can even make things worse as this aid is often distributed initially in the capital. The aid keeps prices low and attracts even more farmers to the city.

A large metropolis requires infrastructure in order to maintain reasonable living conditions. Throughout history large cities had a maximum area, which was determined by the distance that somebody could travel in about an hour by the dominant mode of transport. Transport is crucial for a metropolis, But the transport that is dominant in most metropolitan areas also creates problems—overcrowded roads, traffic accidents, air pollution and noise.

Public health is also an important issue. Because so many people live close together, contagious diseases can spread easily. Adequate sewerage systems and reliable water supply are often lacking and healthcare systems are insufficient.

But perhaps one of the main problems of the large metropolis is a social one. Because many of the people attracted there from rural areas and often from abroad fail to obtain employment, the social structure of the metropolis is often weak. City areas may be controlled by gangs who exist through (violent) economic parasitism and who create a climate in which initiatives that contribute to social and economic development are doomed.

In industrialised nations, transportation has developed to such an extent that inhabitants increasingly seek the absence of noise and the spatial freedom of the countryside. Commuting has grown enormously. Motorways or public transport take commuters to their city job within an hour. Telecommuting means that they need to make this trip only several times a week. Traffic jams or train delays are the major annoyance. This development means that the cities in industrial nations, which were once densely populated areas, tend to grow in size but not in population. Previously undisturbed ecosystems in rural areas and nature reserves near cities are the main victims of this urban sprawl.

From common good to private property[21]

Historically, resources that were once abundant often became increasingly scarce. This created a need for mechanisms to control the distribution of scarce resources. Scarce resources were no longer free.

21   G. Hardin, 'The Tragedy of the Commons', *Science* 162 (1968): 1,243-48.

The early hunters and gatherers did not own any land. Private landownership was unknown to these civilisations (and still is for some nomadic peoples). When they settled, the areas used to grow crops and the livestock became the possessions of families.

For a long period in Europe, the meadows were not privately owned and were called the 'commons'. Livestock belonging to all the villagers could feed on the pasture of the commons and herding was sometimes even a collective activity. Under the pressure from an increasing population, individual families tried to increase the numbers of their livestock and overgrazing ruined the pasture. The solution was to divide the commons into parcels that were privately owned.

Administrative systems evolved for land property rights. This did not happen only to meadows. Similar systems apply to the use of forests—hunting rights and the free collection of wood and fruits were increasingly replaced by private rights. Water became scarce in many regions, with conflicts often developing between upstream and downstream villages regarding the use of water. Such conflicts had to be settled by carefully observed compromises regarding quantities of use and water quality (pollution). This type of conflict occurs even between countries (e.g. the water of the Jordan River [Syria, Israel–Jordan–Palestine] and water of the Euphrates and Tigris Rivers [Turkey–Syria– Iraq]).[22]

The tragedy of the commons suggests that our growing consumption of a scarce natural resource will inevitably lead to private ownership of this common good. Legal and organisational systems have to be created to administer this private ownership. For example, international treaties regulate fisheries and various countries set annual quotas. The privatisation of the Earth's capacity to break down waste is seen in the introduction of tradable emission permits as in the USA. In the Netherlands, cattle farming is regulated by tradable manure rights. The Kyoto Protocol introduced the concept of carbon dioxide (CO_2) emission trading systems such as the European Union Emission Trading Scheme (EU ETS).[23]

Industry, social conflict and alienation

The industrial mode of production brought about an unprecedented level of material well-being—never before have so many people consumed so much material wealth. Not only is the economic system able to provide for our basic needs, it can also provide us with our wants (i.e. various things that are not essential but increase our quality of life and joy from living).

The success of industrial production stems from the trend to turn products into commodities, i.e. products become identical and applicable wherever

22 P.H. Gleick, 'Water, War, and Peace in the Middle East', *Environment* 36.3 (1994): 6-42; www.atmos.washington.edu/2003Q4/211/articles_optional/Water_War_Middle_East.pdf, 2 March 2006.
23 europa.eu.int/comm/environment/climat/emission.htm, 8 November 2005.

you are. A product is no longer made for direct consumption but to be traded to an unknown customer. This division of production and consumption has various consequences. Workers have no idea what happens to their product. For example, the steel they produce might be used to make beds for the local hospital or to build a submarine that is sold to a country on the other side of the world.

Consumers may have no idea where a product comes from. A garment might be produced by exploited children that are kept from school or by collective producers trying to build up a livelihood.

There is often no way of knowing, which makes it difficult to behave responsibly. As a result, it is difficult to bring about moral obligations between producers and consumers. We have become dependent on trademarks and labels to feel responsible towards our fellow citizens.

The division between production and consumption created a new phenomenon—unemployment. Unemployment is an important social problem. It is often a trap from which it is hard to escape—the reduced income and social contacts further decrease the chances of finding a job. The large scale of contemporary industrial production means that the closure of a production site can have a profound impact on the local community. Creating meaningful and lasting employment is therefore important—even in many areas of rich countries.

The larger production units of industry also created more hierarchical organisations. Hierarchy was nothing new but interpersonal relations had previously always balanced it. In Ibieca, for example, there were sharp conflicts between the landless 'casas' and local 'amos', but these conflicts were appeased by the interpersonal relations within the village.

The interpersonal relation as a balancing factor to conflicts of interest is missing in industrial production. This situation was often different in family-owned companies where the owners were the managers (e.g. Ford, Du Pont and Philips) and many smaller ones. The owners of these companies often felt responsible for their workforce. But, with a few exceptions, professional managers have taken over family-owned companies and been given short-term financial targets.[24] The collective actions of trade unions and, in some cases consumer or government pressure, were important in compensating for the social imbalances that frequently occurred.

Various items consumed in the developed world (especially clothes and shoes) are produced by cheap labour in developing countries. To reduce unemployment, developing countries try to attract investors by offering them incentives in so-called export processing zones (EPZs). By 1997, there were 847 EPZs and the number is growing rapidly. EPZs provide jobs in regions with generally high unemployment rates. But, because of this high unemployment, wages are kept low and trade union activities are often

24 A.D. Chandler, *The Visible Hand: The Managerial Revolution in American Business* (Cambridge, MA: Harvard University Press, 1977).

curbed. The workforce in EPZs is generally unskilled and mainly female. In EPZs: 'companies see labour as a cost to contain rather than an asset to develop'.[25]

Globalisation and diversity?

Globalisation leads to more efficient R&D. Fewer competitors develop similar products so there is less wasted effort. Production may also be organised more efficiently. Industries sell their products worldwide. Producers of specific local variants of a product are taken over or are unable to compete with worldwide operating companies that are backed by large R&D facilities. Thus, globalisation leads to a paradox that resembles the biodiversity paradox. The variety of products that you can buy locally has increased, but the variety of products that are produced globally has been greatly reduced.

Consider, for example, television sets. Between 1950 and 1959, there were 102 manufacturers of television sets in the USA alone. Most of them produced a considerable number of different designs, which were mainly sold in North America. Larger European countries had their own manufacturers, but there was some trade between neighbouring countries.[26] Nowadays the market has grown enormously, but the variety of brands has fallen to about 25. At a local TV shop, choice is larger than in the 1950s but it is at least 50% identical to that in any other shop around the world.

The loss of diversity is considerable. The diversity in the products we consume is also part of our cultural heritage. Moreover, should we not value the relationship between production and consumption—to know who made your product, under what conditions and whether they earn a decent living making that product?

A century ago, farmers produced most of the items they consumed. Nowadays it has become virtually impossible to find out where our products come from. We have to trust public officials and companies not to endanger our health.

Products are not just neutral devices. They are connected intimately to a social world in which these products make sense. This implies that exporting products also means exporting lifestyles. This export of lifestyles is often reinforced in the worldwide marketing campaigns that accompany the global trade of branded products. The internet facilitates this worldwide export of the consumer lifestyles of the developed world. It is not our aim to judge the value systems behind this consumer lifestyle. But it is still possible to observe alternative lifestyles in which elements such as solidarity, frugality, introspection, harmony with nature play key roles.

25 International Labour Organisation, 'Export Processing Zones', World of Work 27 (December 1998); www.ilo.org/public/english/bureau/inf/magazine/27/news. htm, 8 November 2005.
26 www.tvhistory.tv, 8 November 2005.

The emergence of conflicts is therefore not surprising. It is the result of the deep penetration into everybody's daily life of the dominant lifestyle of a globalising society.

Structural inequity

In 1820, per capita income in Western Europe was 2.9 times that in Africa. In 1992, Western Europe's per capita income was 13.2 times that of Africa.[27] The UNDP's *Human Development Report 1999* showed that:[28]

- 20% of the global population received 86% of the world income

- The middle 60% of the population received 13% of the world's income

- The remaining 20% of the global population received only 1% of the global income

However, the distribution of incomes between nations is becoming slightly more equal due to the high growth rates of countries such as India and China (5-8% annually in recent years).

Equity is also far from being evenly distributed within nations. Each year, *Forbes* Magazine estimates the assets of the 400 richest US citizens.[29] To belong to this club required at least US$600 million in 2003. In that year, the joint assets of these 400 citizens amounted almost US$1,000 billion. The richest of the rich, Bill Gates, saw his assets decline to only US$46 billion; this amount is about the total annual income of the 50 poorest nations (calculated by purchasing power parity).[30] The same magazine calculated that there were 476 billionaires in the world.

Meanwhile, there are 34.6 million US citizens living in poverty.[31] After a sharp fall in poverty levels in the 1960s, the percentage of poor people in the USA became relatively stable and has now increased again to about 12.1%. The total annual income of these 34.6 million Americans is equivalent to less than a third of the possessions of the members of the USA's 400 richest people.

---

27   UN Development Programme (UNDP), *Human Development Report* 2003 (New York: UNDP): 39; www.undp.org/hdr2003/pdf/hdr03_chapter_2.pdf, 8 November 2005.
28   UN Development Programme (UNDP), *Human Development Report, 1999* (New York: UNDP): 2; hdr.undp.org/reports/global/1999/en/, 8 November 2005.
29   www.forbes.com/richlist, 8 November 2005.
30   www.geographyiq.com/ranking/ranking_GDP_purchasing_power_parity_dall.htm, 8 November 2005.
31   The definition of poverty is somewhat complicated. For one adult, it is an annual income of less than US$9,573. Cf. US Census Bureau; www.census.gov/hhes/www/poverty.html, 8 November 2005.

A number of poor countries are highly indebted. In the early 1980s, rising oil prices created large fortunes for the oil states. At the same time, investment opportunities were limited as the world was suffering from an economic depression. Banks loaned this money to various South and Central American, sub-Saharan African and South Asian nations, often for short-term purposes. But, because the prices of exported goods from these countries (mainly raw materials) did not recover after the economic recession, the interest they had to pay on these loans brought them back in a vicious circle. Their inability to pay meant that either the economy was harmed due to a lack of hard currency to buy crucial equipment with which to sustain production or the prices the population might have had to pay could have sparked a revolution. Countries that postponed interest payments lost the confidence of the banks, with potentially severe consequences for trade.

In 1996, the World Bank initiated its Heavily Indebted Poor Countries (HIPC) programme to relieve the debt of these countries. The programme aims to halve the US$90 billion debt of 33 of the world's poorest countries in 20 years.[32]

It has long been believed that under-developed countries would develop along the same path as Western countries. Technologies would gradually 'trickle down' to the poor and bring these countries the affluence of the industrialised countries. There are a number of reasons why it does not work this way:

- Technology is not the main issue in development. Technology is only part of the transition that has to be made to provide better living conditions

- Developing countries can never go the same way as industrialised ones because there were no industrialised countries when the West built its industry and thus there were no established competitors

- Technologies cannot just be transferred from culture to culture. They embody the culture in which they are produced (this issue is examined in Chapters 5 and 9). Their marketing campaigns appeal to the lifestyle in which the technologies fit. Unless the people of the world all decide to become Western-style consumers, transferring technologies without regard for the cultural fit will produce new conflict

The future knowledge society

What are the consequences of further globalisation for the various regions of the world? Where will the jobs and the production facilities go? Which regions will benefit and which will suffer? Where will R&D take place? These regions will flourish, as they will attract the higher-paid jobs.

32 www.worldbank.org/hipc/about/hipcbr/hipcbr.htm, 8 November 2005.

R&D is a labour-intensive activity. However, labour costs are not the main issue of concern. Successful R&D is about getting a big idea, patenting it and then bringing the product to market. Thus the quality of R&D labour is far more important than the quantity. Hiring extra people is often not enough. R&D needs new ideas and networks of relations with:

- Other scientists working in universities or other laboratories
- Customers that use the product
- Suppliers that may be able to improve the quality of their supplies.

Successful R&D is at the centre of such knowledge networks. For companies deciding the geographical location of their R&D facilities, the primary search is for the links or nodes in these information networks. Large companies want to be part of all the links in knowledge networks. Marketing, production and management then have to be in close contact with the R&D department in order to integrate the latter's results in corporate strategy and to implement strategic decisions in R&D laboratories.

R&D network formation aims to:

- Establish relationships with academics with access to scientific networks and with other R&D institutes
- Interact within the company in order to operate on the basis of the same strategic vision
- Co-operate with customers to understand their needs
- Co-operate with suppliers to see what they might come up with
- Understand the socio-political context of research

The costs of R&D are still growing faster than average income increases. One factor is the growing complexity of technology, which makes innovation more and more an enterprise of collaborating companies. For example, the electronics, car and materials industries are working together on new car equipment. This tendency for co-operation makes it even more important for the individual company to be located at places where partners can come together for meetings.

As a result, R&D facilities are generally concentrated near corporate headquarters in areas of the world where there is already considerable research activity. Those regions that already lag behind in this respect have no chance. Even regions that are marginal within one of the larger economic blocks (USA/NAFTA, EU and Japan) can lose these economic activities.

Can new economic activities be established in the forgotten parts of the world? This will be very hard:

- Setting up new industries takes an increasing amount of capital, which is not available

- Regulations aimed at product or process safety, protecting the environment, improving labour conditions, maintaining trade standards, etc. often make for difficulties in setting up new companies in existing branches of industry. Existing businesses know how to deal with these regulations while new entrants generally have problems dealing with them

- Governments that offer help to new entrants in existing markets might be accused of triggering unfair trading practices

- Established export markets are not open to new entrants

Under current circumstances, only consistent long-term policies in which developing countries focus on specific regions and sectors might be successful.

Questions, discussion and exercises

1. Spend three minutes looking for information on the country of origin of objects around you. How many different countries did you find?

2. Why is the variety of products worldwide decreasing but local variety is increasing?

3. What are the main reasons for urbanisation?

4. Summarise the main advantages and disadvantages of globalisation. Are there options to combine the advantages of a globalised and a non-globalised world?

5. What is the 'population trap'? What strategies could be used to escape it?

4 Sustainable development and economic, social and political structures

Economy and sustainable development are often seen as adversaries. This chapter discusses to what degree sustainable development can be in harmony with economic forces.

First, the chapter considers the general principles of economic organisation and the advent of the market economy. It is argued that every free market needs active government involvement to make the market really work and to guide the market equilibrium in the direction of the common good. Within the framework of this market concept, various measures can be taken to stimulate cleaner products and cleaner production.

The processes of policy-making at national and international levels are then outlined, with particular reference to how to deal with issues that involve technological expertise and how to involve stakeholders.

Finally, the issue of social change is examined. How can individuals change the social patterns in which they are trapped and work at change?

The organisation of production and consumption

Economic organisation

The productive processes in a society can be organised according to the following principles:

- **Market**, i.e. free interactions between consumers and producers of goods and services. Production and consumption depend solely on consumer demand and the supply of products. Producers and consumers are supposed to optimise their individual cost–benefit ratio. Prices are formed by the ratio between supply and demand at the market. Mismatches in the market may lead to great fortunes on the one hand and impoverishment on the other

- **Hierarchy**, i.e. the production and consumption of goods and services (and their price levels) are controlled by a single actor. This single actor is supposed to have no self-interest and steers the economy for the common good, i.e. to prevent exploitation and market misconduct. Individual freedom to engage in market transactions is curbed. Hierarchy was the dominant mode of economic organisation in communist economies and Nazi Germany, but also in the war economies of the allied powers (USA, UK) between 1940 and 1945

- **Networks**, i.e. production, consumption and prices are neither the result of a completely free market nor set by a complete hierarchy. Checks and balances that aim to prevent socially unacceptable outcomes control the market economy. The state guarantees basic social conditions, healthcare and educational facilities, while the citizens accept high levels of taxation to supply these services

These modes of organising the economy have consequences for the political system. The free market is generally regarded to be an economic system that needs no government interference. This is a major misunderstanding. Without the force of law, the free market will often vanish. The protection of the autonomy and integrity of every individual is a precondition for the market to exist. Only governments can guarantee this condition.

Free markets can be manipulated by

- Joint actions of individual actors (e.g. trusts, but also strikes)

- Deliberate actions of large and/or wealthy individuals

- Acts of defamation

Thus, every market economy needs laws and law enforcement to maintain itself. Moreover, market economies might want to protect themselves against the (economic or violent) actions of their neighbours. International trade laws, immigration laws and military protection are therefore also government tasks. Moreover, a really free market is difficult for some products/services; for example, the construction of two railway networks is generally an inefficient solution to guarantee railroad competition. The same applies to electricity, natural gas or drinking water infrastructure.

A main challenge for any government in a **free-market** economy is to maintain its impartiality towards the economic powers it is supposed to control. Any deal between government and specific companies should be transparent. Individuals have great freedom and low taxation. However, minor incidents can ruin individuals. In a free-market economy, dealing with poverty is not a task for the state but for private parties such as churches.

The **hierarchical economy** does not need laws against economic manipulation. In principle, the government needs to control all individual economic activities. But the abuse of this authority in realising absolute control over all citizens is easy. Moreover, experiments with this form of economic organisation were only successful under conditions of war. Under prolonged peacetime conditions, this form of economic organisation was not very successful as mismatches between demand and supply occurred. The basic flaw is the premise that the economy can be controlled centrally and rationally. It is doubtful whether the wisdom and the information for central control of the whole economy can ever be present.

In the **network form of economic organisation**, no actor has absolute power. Public control over negative effects of market economy can more or less be exercised by media attention and the eventual threat of legal actions. Introduction of new laws is generally by consensus and law-making itself is mainly necessary to keep individual economic actors from misconduct. The sizable tasks assigned to the state often make concerted government action a fantasy. Various government departments often have conflicting interests. Many interests in society are institutionalised in powerful organisations, which are interdependent. In this form of economic organisation, the citizen is more protected against individual misfortune. However, policy changes are hard to introduce as the whole system is build on a complex set of institutions and rules, and a complex network of actors.

The advent of the market economy

Many people nowadays regard the free-market economy as a kind of natural system. However, the market economy is actually something that arose during the 19th and 20th centuries.

In medieval society, producers were not free to take decisions regarding production. Major decisions were made by the guild system and, in a town, only members of the guild could practise the particular craft. Those who were members of the guild made joint decisions on prices, production quantities and quality standards. The guild represented the interests of its members to the authorities.

Guilds not only regulated highly skilled crafts (e.g. weaving, production of silverware and construction) but also low-skilled labour such as the loading and unloading of barges and ships. People could not settle themselves freely. There were various trade barriers even within the various small-scale political entities of that time. The guild system and the trade barriers protected craftspeople from strong external fluctuations: unskilled labour could flock into towns, especially in times of problems with agriculture. Such movement could have ruined the craftspeople's business if unprotected by the guilds.

The guilds also made decisions about the acceptability of technologies. In this capacity especially, the guild system gradually became a barrier to progress. In 1776, the Scottish economist Adam Smith published his famous book, *An Inquiry into the Nature and Causes of the Wealth of Nations*.[1] The book formed a powerful academic treatise in favour of free trade, which it argued was in the interest of everybody. Free trade would lead to specialisation of regions/nations into forms of production in which they had a cost advantage. This would increase overall production and so all parties would benefit.

David Ricardo's *Principles of Political Economy*[2] published in 1817 took this point further. He argued that, even if a nation was relatively inefficient and lacking any cost advantage, free trade would be beneficial as it would lead to specialisation in the activity that was the least inefficient (see Table 4.1).

Production cost calculated in hours of labour	Country A		Country B	
	Without trade	With trade	Without trade	With trade
Wine	20	7	5	5
Pig iron	8	8	14	10

Trade would also be attractive if Country B could produce iron at a lower cost than Country B. The resulting Country A trade deficit would soon give rise to a change in currency exchange rates as trade has to be balanced in the long term.

TABLE 4.1 Example: advantages of trade based on the assumption that labour is of equal value and trading costs are 2 per unit

The French Revolution ended the guild system. After its disappearance, industrial production could blossom but a disadvantage was the unprecedented exploitation of labour and child labour in the 19th century. In times of depression, wages could also fall rapidly.

By the end of the 19th century a 'social question' had emerged that led to revolutionary outbreaks. Russia, for example, was reorganised as a centrally

1 www.adamsmith.org/smith/won-index.htm, 8 November 2005.
2 www.econlib.org/library/Mill/mlP.html, 8 November 2005.

planned economy. After World War II, countries such as China, Vietnam and Cuba also developed centralised economic systems. Eastern European countries were forced by the Soviet Union to reorganise their economies, though Yugoslavia developed a centralised economy with a considerable freedom for smaller enterprises.

The market economy had one basic malfunction in that it was subject to cyclic patterns of development: economic progress created higher incomes and profits. This led to higher market demand and even more economic growth. However, economic decline led to falling profits and wages, which led to lower product demand, etc. Particularly in the 1930s, economic recession created enormous unemployment in the major industrial nations. This contributed to the anti-democratic power takeover in countries such as Germany and Spain.

John Maynard Keynes analysed these economic cycles in his book, *The General Theory of Employment, Interest and Money*.[3] His analysis was used to compensate this cycle through anti-cyclic government expenditure; investments in infrastructure had to be made in periods of recession. However, his theory was often abused politically for raising government expenditure. In times of crises, Keynes legitimised high expenditure; in times of economic progress, government tax revenues were high and thus there was no need to cut budgets. The resulting high levels of public debt limited the policy options of governments and made democratic politicians of financially weaker countries vulnerable to political pressures from large financial institutions.

Environment and markets

Figure 4.1 shows the mechanism of the market. The preferences of consumers are shown by the demand curve, which indicates that demand decreases with increasing price levels. The supply curve indicates that supply increases with increasing price levels. The intersection represents the market equilibrium. This shifts as consumer preferences or wealth or production costs (labour costs, interest rates, capital costs, technological efficiency) shift.

The equilibrium will also shift as a result of extra product taxation or price subsidies: in the case of an extra product tax, the supply curve will be higher and a new equilibrium point will be established. In general, the new equilibrium price is lower than the initial price plus tax because demand goes down and competition will increase. In case of a product subsidy, the supply curve goes down. The new equilibrium point is generally higher than the original product price minus subsidies because there is less competition at the supply side.

3   J.M. Keynes, *The General Theory of Employment, Interest and Money* (London: Macmillan, 1936).

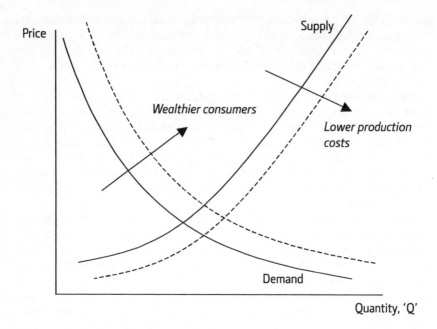

FIGURE 4.1 Changing equilibria on the market

The curves can also be used to show the so-called **rebound** effect: if innovation leads to products with a better environmental performance, they will not be bought if they are much more expensive. However, people are generally prepared to pay a little extra for a good conscience: for example, for eggs from free-range chickens (rather than those housed in cramped chicken batteries).

If the new clean products are less expensive, demand will increase. The fact that such a product is advertised as 'environmentally sound' may appeal to people's consciences and could thus further increase sales.

This occurred, for example, with compact fluorescent light (CFL) bulbs. Although these lights bulbs cost more, they consume 66% less electricity and last ten times as long as incandescent light bulbs.[4] They are also safer than other light bulbs as they produce far less heat. However, once consumers discovered these cost advantages they started using CFLs for additional purposes such as garden lighting—thus displacing some of the environmental gains.[5] The same applies for small cars with lower emissions. For many peo-

4 US Government Energy Star programme; www.energystar.gov/index.cfm?c=cfls.
    pr_cfls, 8 November 2005.
5 In the UK, the Department for Environment, Food and Rural Affairs (Defra) estimated this rebound effect for CFLs to be 30%; www.defra.gov.uk/environment/consult/climatechange/technical/4.htm, 8 November 2005.

ple, the introduction of these cars created an excuse to buy a second vehicle, thus partially eliminating some of the environmental improvement.

Market failure

Hierarchical, centrally planned economies have not been able to produce goods as efficiently as market-based economies. Although the Soviet Union claimed on several occasions in the 1950s and 1960s that its economy would soon catch up with the West, surpass it and provide great wealth for all, it never came close. Shops were often empty. Efficiency of production was low, as was efficiency of resource use. Industrial and mining cities in Russia are still often centres of ecological destruction.[6]

The failure of centrally planned economies to provide for the basic needs of their citizens suggests that the market is an important element in any economic organisation. But, although all participants in the market aim to optimise their individual cost–benefit ratio, the result is not necessarily an optimised situation for everybody. Or, even worse, the optimum may not be reached by the vast majority. The economic cycle is only one factor preventing optimised wealth for everybody.

The market theory presupposes that a consumer's preferences are independent of the decisions of others. Often they are not. Some products, especially high-tech ones, are more attractive if others buy them too (e.g. would it make sense to buy a telephone if nobody else had one?). Individuals are sometimes forced to abandon their preferred product or service as a result of the actions of others. For example, taking children to school by car increases the risk to other children of being involved in an accident near the school. This may stimulate other parents to bring their children by car. The optimum situation might be that no-one brings their child to school by car. The problem is how to reach this optimum. Could the parents reach a workable agreement on the subject?

This is an example of a **prisoner's dilemma**. Free-acting individuals cannot reach the optimum situation unless they can develop some type of joint social arrangement. The following is a classic example of a prisoner's dilemma.

Suppose two criminals are arrested for armed robbery (a crime with a sentence of ten years in jail), but the police do not have a sound case. Without further evidence, it is only possible to convict both men for illegal possession of firearms—a crime with a sentence of only one year behind bars. The police offer parole to each criminal if they help to convict the other. If both co-operate, they will be in jail for five years. For the criminals, the optimum is to keep their mouths shut. But can they trust each other? Consider the alternatives for each one (Table 4.2); not co-operating will bring either ten years

6 See, for example: N.P. Walsh, 'Hell on earth', The Guardian, 18 April 2003; www.guardian.co.uk/g2/story/0,3604,939043,00.html, 8 November 2005.

Prisoner 2 \ Prisoner 1	Co-operates with police		Does not co-operate with police	
Co-operates with police		5		10
	5		0	
Does not co-operate with police		0		1
	10		1	

TABLE 4.2 Prisoner's dilemma

or one year in jail, while co-operating will result in either five years or no imprisonment. Thus it makes sense for each prisoner to co-operate with the police—especially if they are not sure if they can rely on their mate. They can reach their optimum situation only if they trust each other well.

The prisoner's dilemma often occurs in society when we base our actions on assumptions regarding the behaviour of others. For example, government calls to minimise car use during periods of smog are often ineffective as many people see their own contribution as small and do not think that others will stop driving.

The prisoner's dilemma shows us that the free market could lead to an equilibrium that is not necessarily the best outcome for the common good.

Market competition is often possible only by offering suboptimal solutions. For example, airlines often seek to compete with others by offering passengers lower fares if they are willing to make a detour. But this type of market competition does not lead to environmental optimisation as the amount of passenger-miles will increase for the same number of trips.

One of the areas where the market mechanism fails is the provision of **public goods** such as road signs, bridges, parks, public health, clean air and clean water. Public goods are generally characterised by the following:

- Their costs are (almost) independent of their use
- People can rarely be excluded from making use of the goods

Example: air travel

I once met a British colleague at Narita airport in Tokyo. He was travelling home by KLM through Amsterdam airport to London. I was travelling with British Airways through London to Amsterdam. A potential win–win situation (i.e. both travelling directly to our destination) would save each airline one ticket between Amsterdam and London. This was impossible.

Science can also be regarded as a public good when it is not aimed directly at creating new products. The beauty of buildings and cities is a public good—though we might disagree on its value. Non-public organisations such as churches and charities sometimes provide public goods. However, their activities are generally considered an addition to government responsibilities and not a replacement. The provision of public goods requires resources and their quality is therefore often an important political issue.

Markets require that traded products can be owned. However, ownership is not always possible. For example, its very nature means that knowledge is difficult to own; it can generally be replicated with minor costs, especially with the advent of ICT. The knowledge embodied in innovative products is often covered by patents to protect the ownership of this knowledge. Creative knowledge is protected by copyright. But these rights are often infringed without punishment. Moreover, knowledge is often embodied not just in a piece of art or an innovative product. It is also the people within companies that embody the knowledge. When they leave a company, their knowledge is lost—not to society, but to the company that paid for their knowledge creation. This lack of ownership of knowledge causes companies to invest less in knowledge production than the optimum. Governments may therefore decide to use public funding to stimulate the production of knowledge.

In addition, individual parties in the market may find it too expensive or too risky to develop or purchase expensive equipment. For example, farmers may not be able to purchase mechanical equipment for improving production unless they co-operate.

This factor plays an even stronger role in research and development (R&D). Not only are costs enormous, but there is also no guarantee of success. For example, nuclear fusion research or extraterrestrial exploration of rare minerals costs too much and is too risky for any market party to pursue. Governments will need to play a role if we think these technologies should be developed.

The market has little concern for elements of national pride and prestige, which are sometimes paramount in certain activities. For example, the first supersonic civil aircraft Concorde would never have flown and Neil Armstrong would never have made the famous leap for humankind onto the moon. The Olympic Games, as we know them today, would never have existed if governments had not supported them in order to demonstrate national or metropolitan prestige. However poor, any nation starts its own airline to carry its flag around the world.

Sometimes, the market produces outcomes that are considered socially undesirable. This is probably the most politically controversial element of the market. Should people die of famine when there is sufficient food? Should people die due to lack of healthcare just because they cannot pay for a doctor or the medicine that is prescribed? African countries threatened to break international patent treaties to force pharmaceutical companies to sell them HIV medicine at reduced prices. If prices had not been reduced, thousands of people would have died. A number of governments introduced public insurances to guarantee their population minimum healthcare levels. This negative side of the free market (i.e. that some people cannot afford to fulfil basic needs) is one of the key issues dividing the political landscape throughout the world.

In times of unemployment, jobs can be so scarce that wages drop below subsistence levels. The legal minimum wage is therefore a well-known phenomenon in many countries. Similar discussions regarding the outcome of the free market are found for issues such as housing and education. Some countries have rationing systems designed to curb the free market for access to cheaper housing. Basic education is often free and access to higher education is generally determined by learning ability and not by purchasing power.

The decisions of suppliers and customers may also affect third parties. In many densely populated cities, for example, your neighbour's decision to buy a car makes it harder for you to find somewhere to park yours. One solution might be to make parking spaces (a public good) privately owned, but this might involve extra costs for administration and controls to prevent illegal parking.

A final issue in the market economy is that of costs. The market price is supposed to reflect optimised production costs; customers are supposed to prefer best quality for lowest prices. However, some flaws may occur:

- Products may have hidden deficiencies. Customers in their search for best quality could be deceived into buying unsafe or decayed products. The quality level is particularly hard to establish when buying 'know-how' or advice

- Products may require servicing. Once customers have bought a specific product, they often depend on the supplier for service. The supplier may exploit this service monopoly by overcharging. Printer cartridges are a good example: having bought a relative cheap printer, you can find yourself having to buy expensive ink cartridges

- Producers can try to lower their costs by 'externalising' them: for example, by illegally dumping waste. Hiring cheap foreign workers is an example of externalising social costs. The social integration problems that emerge are left for governments. Safety and health risks for workers are sometimes not covered by employers, making the worker pay the price for lower production costs

- The market has its price. Free parking spaces that are paid for by the government do not require meters, traffic wardens and fines, and are thus more efficient for society as a whole. However, the tragedy of the commons (see Chapter 3) teaches us that people tend to consume more resources if they are freely available

The equilibrium of supply and demand reached by the market is not always the optimum (lowest costs) mode of economic production. Changing to more sustainable production may mean that a lower cost level is achieved in the long run—especially if all (external) costs are included.

Free market and emissions

Traditionally, industrial emissions are regulated by permits or licences issued by government regulators. In this way, local residents are protected from too high concentrations of toxic fumes, water contamination or other hazards from an industrial site. A problem arises, however, if the impact of the emissions is not just local but continental (e.g. sulphur dioxide and nitrogen oxides $[NO_x]$) or even global (carbon dioxide $[CO_2]$).

Companies demand to be treated the same way as foreign competitors and therefore want similar emission rights. Hence, the country that allows maximum pollution sets the standard and the resulting level of pollution may be unacceptable. Some companies may find it easier to reduce emissions, with lower costs, than others.

The introduction of tradable emission rights creates a free market for emissions, which can optimise emission reductions. The company that can reduce its emissions at the lowest cost can sell its emission rights to a company that finds it less easy to reduce its emissions. The introduction of trading rights means that emission reductions are not just enforced by licences, but are also stimulated by a profit drive. Companies can therefore start to reduce their emissions as a profitable process. Companies that want to expand have to acquire emission rights; the costs involved may persuade them to invest in technologies with lower emissions.

Tradable emission rights sound a marvellous idea but there are a few problems.

- What will the starting situation be? Will every human being have the same emission rights to start with? Or will the starting situation be the global situation of today in which the pollution of the developed world is legitimised by assigning it emission rights?

- If emission rights are assigned to countries that might reassign them to companies, do we allow international trade? This could lead to poor countries selling emission rights to rich ones, thus creating a barrier to their future development[7]

---

7 See for example: U.E. Simonis, *Internationally Tradeable Emission Certificates: Efficiency and Equity in Linking Environmental Protection with Economic Development* (Berlin: Wissenschaftszentrum Berlin für Sozialforschung GmbH [WZB], Discussion Paper FS-II 96-407, 1996; skylla.wz-berlin.de/pdf/1996/ii96-407.pdf, 8 November 2005).

- Is it possible to establish the necessary technological and administrative procedures? What if countries do not comply with international obligations?

The free market and the depletion of non-renewable resources

Why should depletion of resources be a problem? The market mechanism will lead to higher prices if non-renewable resources become scarcer. This will lead to a larger availability of resources (as more can be exploited profitably). It will also stimulate the development of substitutes for the non-renewable resource (renewable energy for fossil fuels, better recycling for metals, etc.).

Though there is no guarantee that we find solutions, it seems likely that we will manage somehow. But will prices anticipate market shortages to such an extent that there is time to develop improved technologies? Thus:

- Before we run out of non-renewable resources, when will prices start to rise and by how much?

- How long does it take to develop the new technologies that could substitute for the old ones?

Price rise for non-renewables

In general, markets are determined by short-term changes. For example, the oil price increases in the 1970s were due to effective cartel formation by the Organisation of the Petroleum Exporting Countries (OPEC). Oil prices peaked in the early 1980s with the Iraq–Iran war and they were at a historic low in the late 1990s (see Figure 4.2).

Oil prices tend to follow short-term phenomena such as the risk of military conflict in the Middle East, cold winters in the USA, or less oil production due to hurricanes in the Caribbean. Some individuals may even play decisive roles. For example, Saudi Arabia has about a quarter of the world's proven oil reserves.[8] Therefore the Saudi royal family, which rules the country with absolute power, has a major influence on world oil prices and their financial needs, and political motives can be decisive for the oil markets.

Contrary to economic wisdom, countries may increase production when prices drop or decrease production when prices rise in order to maintain a continuous flow of foreign currency to cover their budgets.

---

8 BP statistics; www.bp.com/liveassets/bp_internet/globalbp/STAGING/global_assets/downloads/T/Table_of_proved_oil_reserves_at_end_2002.pdf, 8 November 2005.

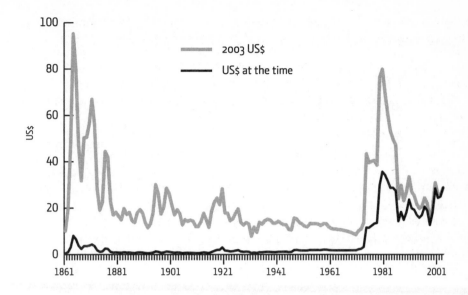

FIGURE 4.2 Crude oil prices calculated in 2003 US$

As yet, metal ore prices are also not determined by the threat of scarcity. Falling demand and the falling-apart of mining cartels have led to falling metal prices.

Prices of raw materials do not reflect a gradual depletion. The threat of global depletion of raw materials has increased since the 1970s, but there is no evidence that this has influenced prices. If market prices anticipate resource depletion, we cannot assume that the warning period will be long enough to give us time for proper solutions.

New technologies as substitutes for the old ones

The oil industry presents a historic example of the effects of rapidly increasing prices on the development of alternatives. When OPEC increased the price of oil fourfold in 1973, a number of things happened.

- Both consumers of oil products and oil producers increased their efforts to find alternatives

- The producers increased their exploration effort dramatically and applied ways to boost oil recovery from previously 'exhausted' or uneconomic wells

- Consumers started substituting oil with other fuels and greatly increased investments in more efficient consumption

Betting on the scarcity of ores

In 1980, economist Julian Simon and Paul Ehrlich (together with John Holdren, the author of the IPAT equation in Chapter 1) disputed the scarcity of ores. Ehrlich had predicted massive shortages within a decade, while Simon claimed that metal ores were abundant.

Simon challenged Ehrlich to put his money where his predictions were. He offered Ehrlich a bet. Ehrlich could pick a quantity of any five metals worth $1,000 in 1980. The winner would receive from the loser the difference from the 1990 price (after adjusting for inflation). Ehrlich agreed to the bet and chose copper, chrome, nickel, tin and tungsten.

By 1990, all five metals were below their inflation-adjusted price level in 1980. So, Ehrlich lost the bet and sent Simon a cheque for $576.07. The reasons for these of lower prices included:

- New mines ended a near monopoly (nickel)
- Technological improvements lowered costs (especially chrome)
- Replacement by plastics and composites
- Fibre-optics replacing copper wires
- Ceramics replacing tungsten in cutting tools
- Tin lost to aluminium in cans

The outcome of the bet shows that metal prices are determined by actual oversupply or short supply on the market.

Source: www.overpopulation.com/faq/People/julian_simon.html, 8 November 2005

As a result of the activities by the oil producers, oil resources increased dramatically. As a result of consumer-side innovation and substitution, oil consumption fell slightly in 1974–75 and did not increase between 1973 and 1986.

In 1972, it was predicted that oil consumption would double by 1982. This implied that oil consumption would be about halved by the technologies developed at the consumer side. But the technologies that were developed were generally easy ones, e.g. switching to natural gas and insulating buildings. Wind energy, which was put forward as alternative source of energy, advanced very slowly. Only some 30 years later is wind energy gradually achieving a market share greater than 1% of energy supply.[9]

9 In 2005 Germany, Spain, Denmark and Portugal produced more than 5% of their electricity by wind power. All other nations produced considerably less electricity by wind power (home.planet.nl/~windsh/stats.html, 17 March 2006).

Thus, price increases in raw materials will probably lead to much extra effort, which may produce extra supply and less consumption. However, these are the easy results achieved mainly by current (optimised) technology. It is unlikely that price increases will produce radically new technological alternatives within short periods of time.

Internalising external costs

To become a winner in free-market competition, market parties not only try to obtain the best bargain but also to shift their costs to others. The environment often suffers by the 'externalisation' of costs promoted by market competition. If those costs were included in the product, consumption would be less and money would be available to compensate or mitigate for the environmental harm they do.

The prices of products should therefore reflect all the costs involved in providing that product. For instance, the social costs of production are the costs of work-related accidents. These costs are frequently not covered by the producer who caused them.

Totalitarian regimes did not compensate for the social costs of their large projects. For example, digging the Belomor Canal connecting St Petersburg to the White Sea resulted in 50,000–100,000 deaths between 1931 and 1933.

However, workers who suffer from work-related hazards are often not fully compensated in free-market states. For example, asbestos workers who developed cancer only recently received some compensation. Some 55,000 people die each year as a result of construction accidents. For example, construction of the Twin Towers in New York (1968–73) resulted in 60 deaths and building the Channel Tunnel between Dover and Calais involved 11 deaths.[10]

Can the loss of life or vital functions be compensated by money? What about work diseases that cannot be directly related to a cause? How can you compensate the extra cancer victims caused by the Chernobyl nuclear meltdown in 1985? How do you differentiate between a Chernobyl victim and someone suffering from 'ordinary' cancer?

It is necessary to attribute these external costs to products and services. However, this is not easy. In practice, the victims are hardly ever compensated and, if so, it is often the government and not the producer that compensates them.

Damage to nature is often hard to express in monetary terms. It might be possible to compensate for CO_2 emissions by planting forests, but what about the extinction of species or the degradation of ecosystems?

Internalisation of costs is important to create economic stimuli for more sustainable practices, but not everything can be compensated financially.

10 T. Zuidema, 'Een dode per miljoen geinvesteerde dollars', *Technisch Weekblad*, 30 January 2004: 7.

Internalisation of costs cannot fully substitute for the effective regulations of occupational health and safety, and environmental protection.

The example of transport

The costs that someone pays for private car transport cover the manufacturing costs of the car and its fuel costs. The tax paid by the owner is intended to create and maintain the necessary associated infrastructure. Car owners are also generally obliged to buy insurance to cover damage that they might inflict on others.

Payments for car insurance and infrastructure are often not related to the actual use of the car. In some countries, the use of longer-distance motorways and expensive structures such as bridges and tunnels are charged to drivers as they actually use them. But, once car owners have paid their fixed charges, they can generally decide to drive more at low additional costs. The costs of car parking are often not covered by the car user. Therefore, many people prefer to park their car in the street, even if they possess a private garage.[11]

Each year, car accidents take the lives of about 885,000 victims worldwide.[12] The number of serious injuries is much higher. If these victims or their relatives are compensated (if at all), the costs are often covered by the state.

Motor vehicles have become a major source of pollution in many cities. Ambient air pollution can pose a significant health risk in cities. Worldwide, about 1.1 billion people in urban areas are exposed to air pollution levels in excess of World Health Organisation (WHO) standards. The problem is particularly pronounced in cities with large numbers of poorly maintained cars. Various studies suggest that air pollution is responsible—primarily through respiratory or cardiovascular diseases—for 2–3% of all deaths.[13] Drivers do not generally pay for damages caused by car exhaust fumes.

Car noise is often regarded as a nuisance and not a cost factor. However, this nuisance has financial consequences, e.g. on the price of real estate. A recent Danish study suggests that, for dwellings exposed to at least 55 dB, their value falls by 0.49% per dB increase in noise level.[14]

Moreover, how can we compensate for the damage to nature? Motorways and other roads are responsible for high death tolls of wildlife. In addition, habitats are cut into small plots by the barriers created by motorways and

11 Cf. The 'tragedy of the commons' described in Chapter 3.
12 World Health Organization (WHO), *World Health Report 1995: Bridging the Gaps* (Geneva: WHO, 1995; www.who.int/whr/1995/en, 8 November 2005): 35.
13 World Resources Institute (WRI), 'Citywide Problems: Ambient Air Pollution'; pubs.wri.org/pubs_content_text.cfm?ContentID=955, 8 November 2005.
14 T.B. Bjørner, J. Kronbak and T. Lundhede, *Valuation of Noise Reduction: Comparing Results from Hedonic Pricing and Contingent Valuation* (Copenhagen: AKF Forlaget, 2003; www.akf.dk/eng2003/noise_reduction.htm, 8 November 2005).

roads. This can be even more devastating to wildlife, especially migrating species, as populations become too small to reproduce. Accidents with fuel and leaks (during refining, storage, from vehicles, etc.) can have a major impact on water quality and marine life. The fuel additive methyl *tert*-butyl ether (MTBE), which replaced toxic lead additives in car fuels, is devastating to groundwater quality. MTBE was detected in 8.7% of US drinking water wells in a survey carried out for the US Geological Society.[15]

At the end of its life, the car becomes waste. Its steel content is often recycled but recycling of its other components is not yet widespread.[16] Landfilling of these materials can lead to soil and water pollution.

A European study[17] estimated that the external costs of transport were about 8% of gross domestic product (GDP). Car use caused the largest share of external costs (58%), followed by heavy-duty vehicles (21%). Road transport as a whole accounted for 92% of external costs while the share of rail and water transport was very small. Passenger cars, trucks and aviation have the highest external costs per transported unit. The shift of transport from rail and public transport towards road and aviation, which is ongoing in the EU, will therefore increase external costs.

How can we include the costs of compensating for or eliminating pollution, and eventually the healthcare and loss of labour costs from increased pollution levels and accidents? Internalisation of costs cannot solve all problems—some things remain unacceptable, no matter how much compensation is paid. However, internalising costs may help to create the stimuli to make transport more sustainable.

- It might stimulate us to travel and transport less by looking for local available alternatives to fulfil our demands

- It might stimulate consumers to choose more sustainable alternatives for transport

- It might stimulate producers to develop innovative transport facilities, which are cheaper when the external costs are included in the price

Hence, flexible pricing instruments are needed to internalise the external costs in an effective way. New ICT is needed if we want to internalise the external costs of transport fully. These technologies should enable us to receive a bill for our transport use according to mode, time of day and route.

---

15 US Geological Survey; sd.water.usgs.gov/nawqa/vocns/Nat_Survey_Summary. pdf, 8 November 2005.

16 In September 2000, the EU adopted the End of Life Vehicle (ELV) Directive, which aimed at increasing end-of-life vehicle recycling of vehicle materials and parts (europa.eu.int/comm/environment/waste/elv_index.htm, 17 March 2006).

17 European Environmental Agency; themes.eea.eu.int/Sectors_and_activities/ transport/indicators/cost/TERM25%2C2002, 8 November 2005.

Moreover, the technology should give clear information to consumers in order to give them a clear choice.

Such advanced 'road pricing' technologies also involve costs:

- Hardware
- Its operation and maintenance
- Cost of abuse
- Enforcement

However, the whole system would become counterproductive if these costs are too high compared with the overall cost. Moreover, an international standard is necessary to prevent cross-border problems.

Policy-making and social change

Technology assessment

An important consequence of the internalisation of all transport costs could be that such billing systems would have to hold data on the transport behaviour of most people. Access to that data can be controlled by law, but privacy will be an important issue. Few people would probably object to the prosecution of criminals as a result of such systems, but what about automatic billing for every speed violation? What about the potential access of intelligence agencies to these data? This could help to track terrorists, but could it also be used to track the government's political opponents?

The study of these kinds of societal consequences of new technologies is called technology assessment. The following definition was proposed at the beginning of the 1970s when it entered the policy arena:

> Technology assessment is an attempt to establish an early warning system to detect, control, and direct technological changes and developments so as to maximise the public good while minimising the public risks.[18]

Technology assessment developed as a new activity for experts advising parliaments.[19] In 1972, the US Congress established the Office of Technology

18 M.J. Cetron and L.W. Connor, 'A Method for Planning and Assessing Technology against Relevant National Goals in Developing Countries', in M.J. Cetron and B. Bartocha (eds.), *The Methodology of Technology Assessment* (New York: Gordon & Breach, 1972).
19 For methodology of the process, see F.B. Wood, 'The Status of Technology Assessment: A View from the Congressional Office of Technology Assessment', *Technological Forecasting and Social Change* 22.3/4 (1982): 211-22.

Assessment (OTA). Similar technology assessment institutes followed in Europe (Table 4.3).[20]

Country/body	Institute
Austria	Institute for Technology Assessment (ITA)
Czech Republic	Prague Institute of Advanced Studies (PIAS)
Denmark	Teknologirådet—The Danish Board of Technology (DBT)
EU	Institute for Prospective Technological Studies (IPTS)
European Parliament	Scientific and Technological Options Assessment (STOA)
Flanders, Belgium	Flemish Institute for Science and Technology Assessment (viWTA)
France	Office Parlementaire d'Évaluation des Choix Scientifiques et Technologiques (OPECST)
Germany	TAB (Büro für Technikfolgen-Abschätzung)
Netherlands	Rathenau Institute
Sweden	International Foundation for Science (IFS)
UK	Parliamentary Office of Science and Technology (POST)

TABLE 4.3 Technology assessment institutes in Europe

Technology assessment institutes often face a dilemma. Should they interfere in 'hot' political debates, with the risk of making enemies, or should they avoid controversial subjects, with the risk of becoming invisible?[21] In the long run, only the latter strategy can be successful as only a good reputation will help the institute to survive political attacks.

In the transport example above, it does not matter particularly whether the government does intend to abuse the systems to disregard privacy of data. What matters is whether citizens trust their governments when they say they will not abuse the data on an individual that comes into their possession.

Internalisation of costs—especially through road pricing—is often seen as an attempt to increase taxation. In principle it is not. It is a different way of raising the money to cover all the costs of transport (provided the system for administering and billing is not too expensive). However, citizens might

20 R.E.H.M. Smits, 'State of the Art of Technology Assessment in Europe'. A report to the 2nd European Congress on Technology Assessment held in Milan, Italy, 14–16 November 1990. European Parliamentary Technology Assessment (EPTA); www.eptanetwork.org/EPTA, 8 November 2005.
21 D. Dickson, *The New Politics of Science* (New York: Pantheon, 1984).

consider it a new way for their government to squeeze money out of their pockets.

Large-scale changes in society, designed to serve the common good, always affect the interests of specific groups in society. Situations in which everybody wins rarely occur. Road pricing will probably affect the overall quantity of transport, thus affecting the interests of the car and oil industries. Public transport organisations may worry about not being able to accommodate new travellers and thus might demand investment in additional capacity.

Financial compensation of minorities that oppose a decision is sometimes justified. But the introduction of new policies would be impossible if all groups that suffer any harm from policy decisions were to compensated.

Policy-making

Ultimately, political democracies are based on majority rule. However, this does not imply that the interests of minority groups can be neglected. If that was the case, then democracies would turn into majority dictatorships. Checks and balances have therefore existed for a long time:

- Majority governments are generally controlled by parliaments with two chambers

- Minority groups can sue the state in courts of law if their interests are not properly considered

- European nationals can ultimately turn to the European Court of Justice in Luxembourg or the European Court of Human Rights in Strasbourg if their national state violates their rights

Some policy issues remain unresolved for many years. Sometimes, they do not even reach the agenda of a country's parliament. But these same issues can suddenly become the focal point of political debate. For example, safety rules are unlikely to be a political issue unless a disaster has taken place. Traffic victims were a political issue in the 1970s, but nowadays seem to be more or less accepted as being the price of mobility. Ideas from the 1970s regarding road pricing seem to be catching on now.

But what makes an idea's time come? The answer is that the right combination of developments at three levels is important to create a 'policy window' in which policy-makers can pick up new subjects and make leaps.

- At the problem level, the issue must be recognised as significant. Various citizens, interest groups and journalists influence this recognition

- The expert level, at which options and alternatives are identified, must be able to generate paths to a solution

- The political level with its dynamics of elections, changes in government and (changes in) political programmes must be able to reconcile the issue with policy. Issues must be consistent with general policies; a government that makes deregulation an issue can find it difficult to deal with a problem that requires the introduction of new regulation

A policy window occurs when there is an opening for new views. This is usually triggered by a major event such as a crisis or a new international agreement. Policy windows provide the opportunity to consider alternative issues and solutions seriously.[22]

An issue often receives attention in the policy arena for a limited time. The former Dutch Minister for the Environment Pieter Winsemius[23] formulated four stages of policy-making:

1. Problem recognition

2. Policy formulation

3. Policy implementation

4. Results evaluation

The problem recognition phase may be very long because the issue does not reach a policy window. If it does, political attention is high in stages 2 and 3. In stage 4, hardly any politician cares about it and the issue is taken care of through normal administrative procedures.

International decision-making

Our world consists of sovereign nation-states. This means that there is no formal authority above the national state. International institutions are generally created by treaties and assigned power by those treaties. National states may decide not to participate in international organisations or to withdraw from them. Eventually, they may refuse to carry out their obligations under those treaties.

The United Nations is the entity in which global problems are discussed. Its General Assembly is the forum where all nations meet. The Security Council discusses crises and threats of armed conflict. The Security Council is made up of five permanent members (USA, UK, France, Russia and China) with the power to veto any decision and ten members elected by the General Assembly.

---

22 J.W. Kingdon, *Agendas, Alternatives and Public Policies* (New York: HarperCollins, 1994).

23 P. Winsemius, *Gast in eigen huis, beschouwingen over milieumanagement* [*A Guest in One's Own House: Contemplations on Environmental Management*] (Alphen aan den Rijn, Netherlands: Samsom, 1986).

The UN and its executive organisations[24] are responsible for co-ordination of global decision-making in many fields, including:

- UN Framework Convention on Climate Change (UNFCCC)[25]
- UN Food and Agricultural Organisation (FAO)[26]
- UN High Commissioner for Refugees (UNHCR)
- UN Development Programme (UNDP)
- UN Educational, Scientific and Cultural Organisation (UNESCO)
- UN Environment Programme (UNEP)
- UN Conference on Trade and Development (UNCTAD)
- World Meteorological Organisation (WMO)
- International Maritime Organisation (IMO)
- International Labour Organisation (ILO)
- International Telecommunications Union (ITU)
- International Monetary Fund (IMF)
- International Court of Justice (ICJ)

Global decision-making is basically a voluntary activity of nations. Although sovereign states are in principle all equal, some are more equal than others. Without the consent of the powerful nations (especially the USA), no global agreements can be made.

The spatial scale of problems is of crucial importance. If spatial scales of causes and effects of problems become larger, the problems become harder to solve. If the spatial scale of effects does not match the spatial scale of the causes of the problem, there will be a conflict of interests as to the responsibility for solving the problem. Causes of problems can often not be controlled entirely by local or national governments. Table 4.4 gives examples of the scale of problems.

Local government action is appropriate for case A. In case B, the international community may compensate the victims. For example, the International Whaling Commission approved quotas for aboriginal whaling hunts (Russian Chukchi, Inuits and Makah Indians) on 22 October 1997.[27] For case C, the international community may offer support to solve the local causes of global problems. This has in fact occurred in relation to protecting the

24 For locations of UN offices and organisations, see: www.un.org/aroundworld/map, 8 November 2005.
25 www.unfccc.int, 8 November 2005.
26 www.fao.org, 8 November 2005.
27 D. Mellgren, 'US meets stiff opposition against Makah whaling', www.makah.com/whaling.htm, 17 March 2006.

| | Cause | |
Effect		Global
	A Soil pollution (e.g. caused by local industrial activities)	*B* Ozone hole above Antarctic Aboriginals being deprived of traditional rights because of international regulation
Global	*C* Nuclear meltdown	*D* Greenhouse effect Depletion of minerals

TABLE 4.4 Scale of causes and effects

rainforest (biodiversity) and increasing the safety of nuclear reactors in Eastern Europe.

Case D is the toughest problem. Which country or organisation will contribute to a solution? Which will play the 'free rider'? Only if there is a sense of international urgency are countries willing to take action. In case of the ozone layer destruction, rapid action on a global scale was possible; Joe Farman and colleagues found proof of the destruction of the ozone layer in 1984[28] and the Montreal Protocol[29] was signed in 1987. As a result, the manufacture and use of CFCs were banned in the industrialised world in 1995 and in developing countries in 2005. Developing countries received aid to install new technologies.

An interesting example is the problem of space debris (i.e. material left after space missions that moves in a permanent orbit around the Earth). Since this problem is a major threat to subsequent space missions, space organisations are willing to enter into agreements that aim to reduce the formation of new debris. Decision-making is relatively simple as the space agencies that suffer from the problem caused it themselves.

Constructive technology assessment

Public decision-making demands solid arguments that demonstrate that decisions are made in the common interest. It also requires people to trust these arguments. The evidence of experts is not enough in modern democracies, as experts are no longer trusted solely because of their status. A transparent and open decision-making process in which interested groups and citizens can participate is important for:

28 J.C. Farman, B.G. Gardiner and J.D. Shanklin, 'Large losses of total ozone in Antarctica reveal seasonal ClO_x/NO_x interaction'. *Nature* 315 (1985): 207-10.
29 www.undp.org/seed/eap/montreal/montreal.htm, 8 November, 2005.

- The democratic quality of society

- The acceptance of decisions by society

Such processes take longer than quick decisions by few officials but may prevent the bitter controversies so often observed in regard to new generic technologies.

Controversies over technology are rarely determined by expert evidence or scientific fact-finding. These controversies deal with the values involved in technology. This does not imply that the experts all agree. For example, physicists can be found arguing for and against nuclear energy.[30]

To bring technology in tune with the demands of society, technology assessment has been reframed as constructive technology assessment (CTA). CTA aims to attune technological developments and social demands. It broadens the decision-making process by involving more stakeholders, including social groups, at an early stage. The term 'constructive' refers to the contribution of activities and studies to the design or 'construction' of new technology instead of previously just criticising new technologies.

Tools for policy-making

Governments have various tools for policy-making, of which the best known are laws. Legislation can make various activities illegal and violation of the law can be punished. Laws are enacted by parliaments, which ensure that they have sufficient support from society. However, laws can take several years to be introduced and are sometimes not the correct solution. Forbid-

Broadening of the decision-making process is important in CTA[31]

Citizens' panels are an interesting way of trying to involve citizens in expert issues. This method was developed in Denmark in the 1980s when 10–20 lay people were invited to make a judgment regarding a controversial technology. They received support and could invite various experts to give testimony before preparing a statement during their final session. The publicity surrounding the panel ensured that other Danish citizens would become aware of the issue.

This method is now seen in many other countries including the USA, UK, Japan and Norway.

30 H. Nowotny, *Kernenergie: Gefahr oder Notwendigkeit?* [*Nuclear Energy: Danger or Necessity?*] (Frankfurt: Suhrkamp, 1979).
31 S. Joss and J. Durant (eds.), Public *Participation in Science: The Role of Consensus Conferences in Europe* (London: Science Museum, 1995).

ding an activity does not make sense if citizens do not have an (attractive) alternative.

Subsidy schemes provide policy-makers with another tool. Subsidies are a means of interfering in the market by lowering the costs of specific products which require promotion to receive adoption, e.g. photovoltaic (PV) installations and wind turbines. Specific taxes increase the price of goods and thus diminish their consumption, e.g. fossil fuel taxes and special taxes on cars.

The consumption of cleaner alternative products can also be stimulated by spreading information or by direct promotion. For example, governments might promote public transport by persuading its citizens that public transport is in the common interest. Spreading information on ways to reduce energy consumption may stimulate citizens to develop innovative energy-saving ideas at home.

An interesting method for government policy-making is to set up (local) experiments to:

- Examine how citizens might change their behaviour

- Find new ways to deal with a new technology

For example, the small-scale ecological communities that have experimented with various forms of ecological farming and energy supply can teach us valuable lessons on how to organise our communities more sustainably. Setting up experiments involving hydrogen cars or electric vehicles would demonstrate:

- How viable the technologies are

- Whether citizens are prepared to adapt their behaviour to use them

- If the technologies can be improved based on experience

Social change

But what can citizens do if the government does not seem to care about their worries? How can citizens who wish to change the society for the better get their point on the agenda?

Suppose you want to change the hierarchy in your university and to have more say about the subjects you are taught. This means breaking the social order in which a student decides which university to attend but the university decides which courses lead to a degree and the professor decides the content of the course. Changes in that system will not be easy—they affect the way we relate to each other and respect each other, and the roles we are supposed to play.

Authority plays an important role in such situations. But how is authority created? Authority is based on the role that a person has within a certain context. Roles are often recognisable by various signs such as:

- The uniforms of policemen and nurses

- The grey hair, speech characteristics and even clothing style of university professors

A famous experiment regarding roles was carried out by Stanley Milgram at Yale University in 1961–62. Volunteers were invited in couples to the university laboratory. One of them was to be a real volunteer and the other an accomplice of the experimenter. The accomplice was to be the test person (student) in a learning experiment and the real volunteer (teacher) was to pose questions and administer electric shocks if the accomplice failed to give the right answer. The experimenter was to record the student's achievements.

However, the real experiment was what electric voltage the teacher was willing to administer to the accomplice. The teacher was instructed to administer an increasing voltage for every wrong answer, starting with 15 volts and going up to 450 volts in 15-volt steps. Each switch had a rating ranging from 'slight shock' to 'danger: severe shock'. The final two switches were labelled 'XXX'. The experimenter instructed the teacher to continue when the teacher expressed doubts. If asked who was to take responsibility for harmful effects, the experimenter accepted all responsibility. No teacher stopped before 300 volts and 65% administered shocks all the way up to 450 volts. Further experiments showed that teachers were less obedient when the experimenter communicated with them via the telephone. Milgram's obedience experiment was replicated by other researchers in Australia, South Africa and several European countries.[32]

The experiment proves the power of authority as no force or threats were used. Although the experiment is now rejected on moral grounds, it offers us a basic insight into social processes. In everyday life, we generally just submit to authority or succumb under their persuasion. Changing our everyday practices means resisting the authorities in our lives.

Thus changing daily practice in a school means changing authority. This often means conflict. There is nothing wrong with conflict as long as people maintain a basic level of communication and respect. The history of humankind's social development is a history of conflict. Conflicts between employers and employees, old people and young people, learners and learned, parents and children, teachers and students seem to be the only way towards emancipation. Sustainable development in the social sense could be described by a paraphrase of Kant from 1786:

> Social sustainable development occurs when people decide to abandon the self-created situation in which nobody listens to them.[33]

32 www.stanleymilgram.com/milgram.php, 8 November 2005.
33 'Aufklärung ist der Ausgang des Menschen aus seiner selbst Verschuldeten Unmündigkeit', in I. Kant, Der Streit der Fakultäte (10 vols.; ed. W. Weischedel; Darmstadt, 1968 [1983]): Vol. 9, 228.

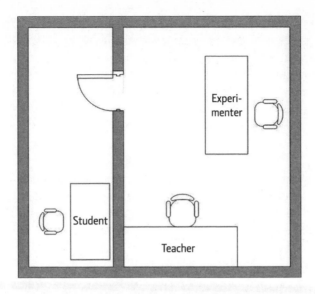

FIGURE 4.3 Experimental setting of Milgram's obedience experiments

Source: www.new-life.net/milgram.htm, 8 November, 2005

Questions, discussion and exercises

1. Table 4.1 shows that trade is attractive only if the difference in production costs is large enough compared with the transport costs. Differences in climate/geography lead to different production costs and thereby trade. Which other factors are affecting trade?

2. Supply cannot always follow demand rapidly. What are the market effects of rapidly growing or declining demands?

3. Discuss the reintroduction of a guild-like economic system. What would be its main advantages and disadvantages?

4. Suppose you had to make a technology assessment of the next generation of civil jet aircrafts. What would be main effects? What would be main differences between the options pursued by Boeing and Airbus (see over)?

Aviation On the very day that marked the centennial anniversary of aviation, Boeing announced, with a great flair for drama, that they are definitively to take the 7E7 Dreamliner into production. *By Mark Plekker*

Haarlem—After a long period of daydreaming about the Sonic Cruiser, the Seattle-based headquarters finally have a feasible plan on the drawing board. At first sight, the 7E7 appears conventional—mid-sized, twin-engined, a non-stop range of 9,500 miles and a passenger capacity of between 200 and 250. The novelty of the design lies in the choice of materials and the efficiency of both the engines and the production.

More than half of the construction of the aeroplane, which currently exists solely in the form of a 7-foot model for wind tunnel testing, will consist of composites. Along with significantly more efficient engines and improved aerodynamic properties, this should lead to a fuel consumption reduction of more than 20%. The use of composites and the resulting weight loss only reduces the fuel consumption by 3%. The rest of the reduction is accounted for by the improved engine efficiency and better aerodynamics.

Boeing's choice for the mid-sized 7E7 also reveals a principally different view compared with Airbus regarding the future development of civil air transport. Airbus believes strongly in central airport hubs and is developing large aeroplanes to fit this purpose. Boeing has staked its design on the growth of mid-sized airports which offer passengers more, but smaller, aeroplanes.

Source: translated from *Technisch Weekblad*, 9 January 2004

5 Technology: the culprit or the saviour?

Many critics have argued that technology is the root cause of the lack of sustainability in society.[1] Their argument assigns the key role in the development of the problems of the modern world to technology:

> People's lives become more and more dominated by technology, without offering the possibility of influencing it. Therefore, technology should be halted in order to return to a more natural world. Every attempt to spur innovation for sustainable technologies is doomed to futility, or could even make things worse as technological domination will increase and (unforeseen) side effects of technology will exacerbate the situation.

However, the history of technology also presents another picture: the social world is not dominated just by technology but is also actively engaged in shaping new technology. Technologies were shaped in conflicts such as those between workers and employers, and reflected the moral, social and political orders of the time.

We need a leap in the environmental efficiency of our production of goods and services. Technological change must therefore be at the heart of sustainable development. This chapter explores the role of technology in society and its relation to science—as science and technology have become strongly interrelated.

---

1 For example, the philosophers Habermas and Marcuse. Also more recently: E. Braun, *Futile Progress: Technology's Empty Promise* (London: Earthscan, 1995) and K. Sale, *Rebels Against the Future: The Luddites and their War on the Industrial Revolution* (Reading, MA: Addison Wesley, 1995).

The most profound changes in the way we deal with science and technology took place over the past 500 years. We focus on this period in the opening section and then offer some theoretical reflections.

Technology and science in human history

Medieval technology

Humans have developed various technological means to support themselves. In the Middle Ages, technological 'progress' diffused only very slowly. The case of papermaking (Figure 5.1) serves as an illustration.

> The basic features of modern papermaking technology were created in China in the early 2nd century AD. In the early 7th century, papermaking reached Korea and Japan. In the 8th century, it reached Samarkand in Central Asia. The technology was established in the Muslim world in the 10th century and in the 11th and 12th in Christian Europe, where it was initially denounced as a technology of the Muslim foe. This attitude remained essentially unchanged until the introduction of the printing press in the 15th century, when paper finally became accepted in Western Europe.[2]

Innovations that would today reach every corner of the world within a few years spread only very slowly, if at all, in medieval times. This is only partly due to slower methods of transport and limited communications. Neither was the slow pace of technological change due to a lack of individual inventiveness. The Christian resistance to papermaking illustrates another feature of medieval society. In a society that sought its solace beyond earthly life, technological change was considered relatively unimportant. European medieval thinking on how nature and society were ordered stressed the importance of objects and people remaining in their 'allotted place'. Financial prosperity through technological innovation was not encouraged. As long as new technology could be absorbed within existing institutions, it was more or less tolerated—being sought actively only in times of sheer despair such as starvation, conquest or flooding.

Technology was part of the world of craftsmen. Craftsmen were generally free men living in medieval towns where they were organised in guilds that protected their interests (see Chapter 4). To be able to execute their craft, they often needed a licence from the local lord. Craft workers generally took great pride in their craft, becoming apprentices at the age of 11 or 12. This apprenticeship lasted for about seven years.

2 Paperonline; www.paperonline.org/index.html, 9 November 2005.

FIGURE 5.1
Medieval
papermaking

Source: www.st-armand.com/Images/200412_CoursPapier.gif, 9 November 2005

Science within medieval Christianity had a very different background. It encompassed the search for truth as a means of better understanding God's creation. It had nothing to do with pure curiosity or with the wish to improve the destiny of humankind. Philosophy was directed towards finding proof of God's existence and solving apparent inconsistencies, e.g. between God's perfection and the apparent imperfections in the universe. The study of law was aimed at reconciling commercial practice and divine law. Sciences that we regard today as empirical and experimental such as botany, physics, astronomy and physiology were part of philosophy. They were directed towards explaining God's divine works and not to practical utilisation.

The ordinary craft workers had nothing in common with this science. Such knowledge was of no value to them. The scientists belonged to the social world of the rulers of that time, i.e. the nobility, clergy or successful merchants.

As explained in Chapter 4, medieval competition was stifled by the structure of economic production, i.e. agricultural production by a peasantry dependent on the nobility and production of a narrow range of goods by craft workers organised in a guild system. The general result was the maintaining of tradition rather than spurring innovation for a 'better' future.

Technological change did occur, but mainly as a result of outside influence and in those societies where the old 'divine order' was challenged most. Thus, Crusaders probably brought the windmill (Figure 5.2) to Europe in the early 13th century. The windmill was of great advantage in lowlands that

Figure 5.2 Medieval windmill type (note the straight wings)

Source: Diderot et d'Alembert, *Recueil de Planches sur les Sciences, les Arts Liberaux et les Arts Méchanique, avec leur explication* (Paris, 1762)

lacked water power. But the technology diffused only slowly because of its impact on social structure, which upset the nobility's traditional milling rights. In the north-west European lowlands, however, the windmill became an increasingly popular and effective means of pumping water. With sea levels rising and agricultural peatlands setting after centuries of use, these countries were in dire need of pumping power. In these areas, windmill-driven pumps were not introduced primarily to expand economic activity but to protect agriculture from deteriorating natural conditions. Later on, population growth prompted a need for agricultural expansion and, with most arable land already in use, windmills opened the way for 'reclaiming' new land.

The Renaissance

European medieval science was rather backward in comparison with the scholarly works of the Islamic world of that time. The classic Greek works were studied in the Islamic centres of learning but hardly at all in Europe. Physiology, philosophy and mathematics reached European scholars particularly through interaction with the Spanish centres of Islamic learning, which were conquered in the 14th and 15th centuries during the Reconquista.

The study of the divine order became more and more empirically oriented. For example, the Danish nobleman Tycho Brahe became famous for his detailed astronomical observations.[3] Inconsistencies appeared and scholars dared to propose interpretations of their own which were not in accordance with the interpretations of the church. For example, the Polish scholar Nicholas Copernicus proposed a heliocentric cosmology.[4] Galileo Galilei's defence of this system culminated in his well-known clash with the Pope: in 1633, the Inquisition convicted Galileo for his heretical theories.

It is not by chance that this clash coincided with the great religious divide within European Christianity. The values that were underlying this split—empiricism, rationality and individual conviction—divided the institutions of Christianity and remained points for debate within the churches.

The changes that took place in the Renaissance (14th–16th centuries) had a tremendous influence on the social climate for technology. Society was more geared towards the individual, who could try to obtain power, property and status. The striving of intellectuals for individual expression included the active design of new machines; we know this, for example, from the extensive drawings of Leonardo da Vinci.[5]

The changes also fuelled a drive to discover the world, facilitated particularly by new, superior techniques of shipbuilding and navigation. This led to a series of famous discoveries such as the discovery of:

- America by Columbus in 1492

- The sea route from Europe to India by the Portuguese Vasco da Gama in 1497

- The first tour around the world by Ferdinand Magelhaes[6] in 1521

An early example of the application of science to technology can be seen in Galileo's work around the turn of the 17th century. Galileo was the first to

3 es.rice.edu/ES/humsoc/Galileo/People/tycho_brahe.html, 9 November 2005.
4 *De revolutionibus orbium coelestium* was published in 1543 just before Copernicus died.
5 The machines of Leonardo da Vinci can be seen in the National Museum of Science and Technology, Milan, Italy (www.museoscienza.org/english/leonardo/invenzioni.html, 17 March 2006) and the Chateau du Clos Lucé, Parc Leonardo da Vinci, Amboise, France (www.vinci-closluce.com/machines.htm, 17 March 2006).
6 Magelhaes died during this tour, but one of his five ships returned safely to Spain.

show that, if air resistance is disregarded, the acceleration of a falling body is independent of its mass. This allowed him to calculate the parabolic orbits of projectiles. The resulting tables proved to be a powerful tool for gunners, enabling them to determine accurately the firing angle needed to hit any target at a given range.

The new empirical science reached a peak in Isaac Newton's mechanical theory. Newton's theories were able to explain planetary motion and terrestrial movements using the same set of equations. Newton and Leibniz independently developed the calculus[7]—a powerful mathematical tool for every modern engineer.

Newton's theories helped in navigation and the development of new instruments for navigators, helping Britain to achieve maritime domination. However, science and technology were 'living apart together' in the 17th and 18th centuries.[8]

Most technology remained predominantly based on traditions that were passed on by apprenticeship. The universities did not have a dominant role in the scientific changes that led to Newtonian mechanics; they remained dominated by classical scholars and frequented by students from the nobility and landed gentry.

Scientific and technological progress

In the 18th century, it was still rare for the results of scientific experiments to be exploited commercially in the form of new or improved products and processes. Technological changes that could, in principle, be calculated by mechanical theories were developed mainly by practical craftspeople without any theory. Although Dutch windmill-builders improved their designs during the 17th century, e.g. by adjusting the angle of the windmill sails along the shaft of the sail (Figures 5.3 and 5.4), their efforts were not based on scientific principles.

It was not until 1759 that John Smeaton corroborated the value of the 17th-century changes in windmill design when he presented results of the first scientific experiments on windmill sails in his *On the Construction and Effects of Windmill Sails.*[9]

A number of key technologies were created long before their underlying scientific principles were formulated. For example, the invention of the first

---

7 British and continental scientists have long argued on priority claims for this invention.
8 R.S. Westfall, *The Construction of Modern Science* (Cambridge/New York/Melbourne: Cambridge University Press, 1971); T.S. Kuhn, *The Copernican Revolution: Planetary Astronomy in the Development of Western Thought* (Cambridge, MA: Harvard University Press, 1957).
9 John Smeaton, *On the Construction and Effects of Windmill Sails: An Experimental Study Concerning the Natural Powers of Water and Wind* (Philosophical Transactions of the Royal Society of London, 51; 1759): Pt 1, 138-74.

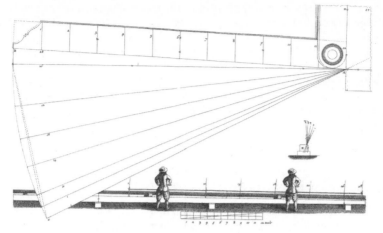

FIGURE 5.3
Drawing showing the angles of a sail on a wing shaft

Source: *Groot Volkomen Molenboek* (Amsterdam, 1734)

FIGURE 5.4 Wing of a windmill (the changing angle of the sail to the shaft is clearly visible)

Source: home.wanadoo.nl/flikkenschild/laeker/molenwiek79.htm, 9 November 2005; © Erik Flikkenschild, Oegstgeest

steam engines[10] preceded the formulation of the thermodynamic principles explaining their operation by more than a century.[11] As L.J. Henderson penetratingly wrote in 1917:

10 By Newcomen in 1712 (see Chapter 3) and Watt in 1770.
11 By Carnot in 1824 and Joule in 1845.

> Science is infinitely more indebted to the steam engine, than
> is the steam engine to science.[12]

The Enlightenment (18th century) can be seen as the closing phase of this changing relation between science and technology. Its basic philosophy of rationalising society implied that technology should be freed of its guild-based rules and traditions of craftsmanship. The French philosopher Condorcet (1743–94) introduced the idea of 'progress'.

- In science, it implied an improved understanding on the causes of nature

- In technology, it implied constructing more and better artefacts to improve human domination over nature

For us, this might sound pretty obvious. However, for 18th-century Christianity and other religions, the idea of seeking progress in life was strange. Rewards for conduct were only to be received *post mortem*.[13]

Science was moving only very gradually away from religious dogma. Before Charles Darwin published *On the Origin of Species* in 1859, biologists based their science on theological assumptions. A good example is the answer to the question how old is the Earth: in 1800, Archbishop James Ussher established the year of origin as 4004 BC.[14]

Before the 19th century, the process by which technical innovation came about was generally one of trial and error. As failures were often dramatic, there was a strong reluctance to change technologies. Churches, for example, were constructed according to rules of thumb that had remained unchanged for centuries.[15] Discovering that a construction would be unstable was a disaster given the enormous investments that medieval cities made in building these structures. A notable example of this is the bell tower of the cathedral in Pisa, Italy (Figure 5.5).

The nation-states that took shape in the 18th and 19th centuries transformed European economies, shifting their basis from local autarchy to

---

12 L.J. Henderson was the first president of the History of Science Society. See: L.J. Henderson, *Henderson on the Social System: Selected Writings* (Chicago: University of Chicago Press, 1970).

13 Christians could deserve heaven by good conduct or remorse of sin. Hindus could reincarnate as a higher being.

14 D. Nelkin, *The Creation Controversy: Science or Scripture in the School* (New York: W.W. Norton, 1982): 25-26.

15 Though modern technology sometimes pretends that its rigour is completely proven in advance by scientific means, unexpected problems can still arise. Bridge construction, for example, is still in part trial and error, sometimes leading to unforeseeable failure. In 1996, the city of Rotterdam had to close its new Erasmus Bridge because of oscillations set up by the wind in the suspension cables. Similarly, pedestrian-induced oscillation in 2000 of the London Millennium Bridge necessitated its closure and substantial reconstruction.

Figure 5.5 **Leaning Tower of Pisa**

national and imperial trade. This transformation required a variety of civil works such as:

- Railways
- Canals
- Roads
- Bridges
- Water control and management systems

In addition, machinery and factories had to be designed to supply large quantities of goods efficiently.

These technologies were not designed according to local circumstances (the practice of the medieval craft workers), but were designed using the universal rules of science. The substitution of tradition by rationality promoted in the 19th century led to a need for a new profession—the engineer.

Engineers were not traditionally trained craft workers but people who were trained scientifically to design and operate technologies according to rationalist principles. For example, the 1847 founders of the Dutch Royal Institute of Engineers[16] explicitly required its members to be trained scientifically.

Engineering schools were first founded in France, e.g. the École des Mines in 1783.[17] Later in the 19th century, engineering schools were established throughout Europe and the USA. As generally the only predecessors of the new engineers were the military engineers engaged in artillery and the construction of fortresses, the new engineers were called 'civil engineers'.

By the end of the 19th century, the 'social issue' dominated the public debate: industrial production had created an urban proletariat and sharp social conflict. Many engineers were sympathetic to political visions in which the state was assigned a leading role in advancing social justice. For a profession that had gained its status by rationalising infrastructure and production, rationalist planning of economic activities was a highly attractive proposition. An example of thinking in terms of industrial progress is shown in Figure 5.6.

This also was evident in the 1920s and 1930s when various sources of energy (electricity, coal, oil, town gas) became available. Many engineers started arguing for rational (state) planning of these energy sources as the competition between them was often creating inefficiencies in infrastructures. In general, the engineering profession of the first half of the 20th century was committed to solving society's problems through rationalised planning. The progress of engineering was well fitted to the progress for which the socialist movement, in particular, was striving for.

Techno-science

Although engineers were scientifically educated, academic scientists did not regard them as colleagues. Polytechnic colleges and institutes were often not allowed to grant doctorates and their social status was much lower than that of universities. However, this view changed as the polytechnics often trained excellent scientists, e.g. Albert Einstein graduated in 1900 from the Eidgenössische Technische Hochschule (Swiss Federal Institute of Technology) in Zurich. Universities in the 19th century concentrated on the classics. The natural sciences generally dealt with phenomena that were interesting, but without any practical use. Industry and trade usually had little interest in the scholarly activities of universities.

This changed first in chemistry. In Germany, chemists developed processes to manufacture synthetic dyes based on coal tar. Bayer AG in Leverkusen was the first company to develop a research laboratory for the scientific study of chemistry in order to create new chemical products. How to do research and how to organise it within the context of industry was itself an

16 Koninklijk Instituut van Ingenieurs (KivI); www.kivi.nl, 9 November 2005.
17 www.ensmp.fr/Fr/ENSMP/Histoire/histoire.html, 9 November 2005.

Figure 5.6 Stained glass window at the Industrial School of Terrassa (Spain) constructed in 1902

Source: *L'escola Industrial de Terrassa 1902-2002, cent anys de vida universitaria* (Lourdes Plan I Campderros, 2002)

innovation.[18] Competing chemical companies soon followed Bayer's example.

By 1890, the electrical inventions of Thomas Edison led to the birth of the electric industry. Research was needed to provide improvements and the General Electric Research Laboratory in Schenectady, New York, emerged. This laboratory acquired fame following the award of the Nobel Prize in chemistry to Irving Langmuir in 1932.

18   G. Meyer-Thurow, 'The Industrialisation of Invention: A Case Study from the German Chemical Industry', *ISIS* 73.268 (1982): 363-81.

The creation of nylon

At the end of 1926, Du Pont decided to undertake a fundamental research programme. One of its main proponents, Charles M.A. Stine, later explained the motives:

> Fundamental research assists one to predict the course of development of chemical industry. Pioneering applied research enables one to achieve certain objectives indicated by fundamental research. Therefore, the continued growth (as distinct from mere expansion) of the chemical industry is dependent upon fundamental research. That is the basic philosophy of fundamental research.

Stine stated that fundamental research also improved industry–university interaction and created consulting specialists for applied research within the company. The main difference to university research would be that:

> In university research, the discovery is a sufficient objective in itself . . .

Du Pont recruited academic scientists and gave them the freedom to engage in the subjects they thought could be useful. In 1928, Wallace Hume Carothers (Figure 5.7), a chemist from Harvard University, was enlisted as head of the organic chemistry group. For a long time, Carothers was doubtful about taking this step because industrial research was not valued highly by academics. At Du Pont's Experimental Station, he started research on the macromolecular concept of polymers—a subject of great

Figure 5.7 **Wallace Hume Carothers**

Courtesy: Hagley Museum and Library

interest and a focal point of debate in chemistry at that time. Carothers made new, long-chain polymers by carrying out well-understood chemical reactions. A flood of publications emerged from this research project that demonstrated that polymers were just ordinary molecules, only longer.

Initially, Carothers was supposed only to carry out fundamental research and Du Pont encouraged the publication of his results. However, new managers in 1930 urged Carothers to focus his research on developing new products based on polymers. He did not resist this pressure. His main concern was the freedom to publish. This was granted to him if he was willing to give the company time to patent applications for his discoveries.

In September 1931, Carothers announced the possibility of obtaining useful fibres from strictly synthetic materials. In July 1935, Du Pont chose to commercialise polyamide-6,6 because the raw materials were comparatively cheap and, at the end of 1937, it was produced in a pilot plant.

The technological problems of production were enormous. The required purity of raw materials was unprecedented and the spinning process differed considerably from conventional processes. About 230 engineers worked on the project and more than 200 patents were granted just for the technological work.

At the end of 1938, Du Pont launched 'nylon'. It was an overwhelming success on the market. Du Pont could not fulfil demand during World War II due to the military demand for nylon parachutes. Nylon stockings disappeared from the market only to return with even greater success after 1945.

The spirit in which Carothers led his team was described by one of the members of his group, the later Nobel Prize winner Paul Flory:

> His approach to science was motivated by boundless curiosity; it was not fettered by superficial boundaries between specialties.

Carothers did not live to see the success of nylon. He committed suicide in April 1937, deeply depressed and, although the first industrial chemist to be admitted as a member of the National Academy of Science, convinced of having failed as a scientist.

Scientists such as Carothers bridged the gap between academics and industrial technologists. As Harvard president James B. Conant said of Carothers' acceptance of a position at Du Pont:

> . . . he had facilities for carrying on his research on a scale that would be difficult or impossible to duplicate in most university laboratories.

Source: somewhat abbreviated from K.F. Mulder, 'Replacing Nature: The Arising of Polymer Science and Synthetic Fibre Technology' in B. Gremmen (ed.), The Interaction between Technology and Science (Studies in Technology and Science, Vol. 3; Wageningen, the Netherlands: Wageningen Agricultural University, 1992): 239-62

The Du Pont Corporation of Wilmington, Delaware, in the USA set up a chemical research laboratory. Du Pont made a fortune during World War I following the invention of its smokeless gunpowder during the late 19th century. Du Pont decided to continue this successful strategy by carrying out even more fundamental research. However, the frictions between academic and industrial research were still partly unresolved.

The history of nylon discovery at Du Pont serves as an interesting example of techno-science in the making.

Science has proved its value to industrial and military interests. Scientific research became the first stage of the development of new products and processes.

During World War II, physicists in the USA suspected that Hitler was working on a German nuclear bomb. They convinced Einstein to ask US President Roosevelt to set up a research project to also develop a nuclear bomb. In the Manhattan Project, top physicists were gathered to build such a bomb. The first ever nuclear explosion took place on 16 July 1945 at the Trinity test site in the Alamogordo Desert, New Mexico, USA (Figure 5.8).

The Japanese cities of Hiroshima and Nagasaki were the first targets of nuclear attacks on 6 and 9 August 1945 respectively.[19] Physics had lost its innocence.

Nowadays, the natural sciences not only contribute to technology. Technology also plays a major role in carrying out science: supercolliders in particle physics, the Hubble space telescope, supercomputers and high-resolution microscopes are technologies without which modern science would be impossible.

Technocracy

New problems emerged in the industrial market economies in the second half of the 20th century. With large parts of society effectively rationalised, many people felt alienated by the complexity of the technological systems that they were part of. Industrial production was often organised according to Frederick W. Taylor's 'scientific management' methods, which implied that production was cut into small tasks that were assigned to a single labourer. The tasks of the labourer were measured and goals for his production were set. In 1913, Henry Ford was the first to combine Taylor's approach with the moving assembly line to produce the Ford Model T. In this way, an unskilled labourer, making just exact copies of one specific part, replaced the craftsman that made every part to fit.

19 There have been many discussions about whether the bombing of these two cities could have been prevented by the demonstration of a nuclear bomb to the Japanese. For the role and views of scientists: see the short biography of Robert Oppenheimer, director of the Manhattan project (www.pbs.org/wgbh/aso/databank/entries/baoppe.html, November 2005) or R. Jungk, *Brighter than a Thousand Suns: A Personal History of the Atomic Scientists* (New York: Harcourt, Brace & Co., 1958).

FIGURE 5.8
Trinity test site—
12 seconds after
the first nuclear
explosion on 15
July 1945

Source: www.lanl.gov/orgs/pa/photos/historical.html, 9 November 2005; © The Regents of
the University of California, 1998–2003

Vast extensions of cities were needed to supply the growing urban population with housing and to improve the living conditions of the lower classes. Rationalised spatial planning provided for efficient use of indoor and outdoor space. However, this method of spatial planning led to a heavy-handed uniformity of the built environment.

But criticism grew from the 1960s. Workers and their trade unions criticised the stupefying character of assembly-line labour.[20] Environmental pollution, ecological destruction and resource depletion became increasingly evident and were deemed to be 'the price of progress'. The progress in military technology led to the creation of weapons of mass destruction, which even created the risk that humankind could extirpate itself.

Although initially specific technological products were criticised for their detrimental impact, 'technology' itself was soon under attack. 'Technological thinking' was accused of being based on a principle of domination over both nature and people. Western societies were accused of being 'technocracies', i.e. governed by expert knowledge.

The technocratic attitude was condemned as:

● Undemocratic in that it sought to control people rather than interact with them

---

20   Charlie Chaplin's film *Modern Times* is a beautiful criticism of the assembly line.

- Short-sighted in that intensive and exponentially growing use of natural resources and production of waste and emissions were bound to end in catastrophe

Engineers and their technologies were not only blamed for creating these substantial problems, but were also more generally accused for harming society. This led to engineers becoming far less committed to social issues and, instead of the social engagement of the first half of the 20th century, engineers tended to retreat from society in the second half. Many engineers regarded politics as irrational and politicians as only seeking their own re-election instead of the common good.

The controversy on nuclear energy that took place in almost every industrial country starting from the beginning of the 1970s exemplifies this perspective. To provide for the projected growing demand for energy, there was no other option then to build nuclear power plants. This progress came at the expense of:

- The unsolved problem of nuclear waste
- The risk of unprecedented accidents
- The risk of spreading the technology for manufacturing nuclear arms
- Threats to liberty caused by measures to counter risks of attack on nuclear installations

Large numbers of people did not agree with this sacrifice. Meanwhile, it turned out that the growth in energy consumption in most countries was not as large as had been anticipated. The controversy that emerged came to an end with the meltdown of the Chernobyl nuclear reactor in 1986 (Figure

FIGURE 5.9 **Chernobyl reactor**

5.9).[21] The technocratic argument that there was only one solution to provide society with energy had failed.

Having grown used to criticism from environmentalists, engineers did not jump on the issue of sustainability when it was introduced by the Brundtland Commission in 1987 (see Chapter 1). Although sustainable development might equally well have been interpreted as a new challenge to the engineering community, many engineers regarded the notion as suspicious and were concerned that it might mean a new attack on their profession.

Over the past decade, there have been some promising initiatives to once again harness engineers and their skills in addressing the challenges facing society such as sustainable development. Will engineers be able to contribute to sustainable development but avoid the pitfall of technocratic domination of nature and people?

Below we examine concepts regarding the relation between science, technology and society.

What drives technological change?

Many people regard technology as self-evident. It is just there and it is the most efficient or convenient way to fulfil a need. But why was it developed? Was it a stroke of luck, of ingenuity, or was it determined by the state of the economy? Can outsiders or even technologists themselves influence the future course of technology towards sustainability? Or are these attempts futile, as the course of technological change is beyond anyone's control?

Technological determinism

One of the most outspoken critics of modern technology was the French philosopher Jacques Ellul (1912–94). Adopting a historical perspective, Ellul sought to understand how medieval technologies differed from the technologies spawned by the rationalist attitudes of 20th-century engineers. In his view, they did so in a number of salient respects. Medieval technologies were:

- Restricted in their sphere of application—technologies were often based on specific local resources and therefore rarely transferable

- Dependent on limited resources and on well-developed skills such as making and repairing tools, or judging weather conditions and tides

21 See, for example: www.chernobyl.co.uk, 9 November 2005.

- Local in character—technological solutions to specific problems were embedded in local culture and tradition

Together, this meant that individuals and local communities were able to influence and shape the technologies they applied. There was, in other words, a degree of technological choice.

According to Ellul's analysis, modern technology has quite a different set of characteristics:

- Automatism. There is just one 'best' way to solve each particular problem and the accompanying technology appears to be compelling across the planet

- Self-perpetuation. New technologies reinforce the growth of other technologies, leading to exponential growth

- Indivisibility. The technological way of life must be accepted completely, with all its upsides and downsides

- Cohesion. Technologies used in disparate areas have much in common

- Universalism. Technologies are omnipresent, both geographically and qualitatively

For Ellul, this meant that modern technology destroys human freedom. In his view, the future of humankind was bleak and there was no way back.

Some of Ellul's basic lines of argument can be recognised in the 'Unabomber Manifesto'. In the 1980s and early 1990s, Theodore Kaczynski—a mathematician and terrorist known as the Unabomber—launched a series of attacks on US airlines and research institutes. In his Unabomber Manifesto, he explained that it was possible for individuals to participate in modern society only if they accepted its technologies. Although these technologies were often legitimised as creating more freedom, the Unabomber argued that they in fact took away more and more of our freedom because they proved to be increasingly compelling for the individual. In the vision of the Unabomber, this would ultimately lead to the destruction of human freedom. And in his view, too, there was no alternative and no way back.[22]

But, in their analysis, Ellul and the Unabomber fail to acknowledge the tremendous advantages that technology has brought us. The number of people living a relatively good life on planet Earth (i.e. well nourished and healthy) is historically unprecedented. Criticising the alienation created by technology as Ellul and the Unabomber did is one-sided, to say the least, and therefore leads to unjustified conclusions. We indeed sacrificed some liberties but, in return, received new ones. Moreover, the debate on nuclear energy proves that the course of technology is not a path leading to an inevitable destiny. Social preferences will be reflected in technologies, espe-

22 www.thecourier.com/manifest.htm, 10 March 2006.

Unabomber on technology: Unabomber Manifesto, paragraph 127

A technological advance that appears not to threaten freedom often turns out to threaten it very seriously later on. For example, consider motorised transport. A walking man could formerly go where he pleased, go at his own pace without observing any traffic regulations, and was independent of technological support systems. When motor vehicles were introduced, they appeared to increase man's freedom. They took no freedom away from the walking man, no-one had to have an automobile if he didn't want one, and anyone who did choose to buy an automobile could travel much faster than the walking man. But the introduction of motorised transport soon changed society in such a way as to greatly restrict man's freedom of locomotion. When automobiles became numerous, it became necessary to regulate their use extensively. In a car, especially in densely populated areas, one cannot just go where one likes at one's own pace; one's movement is governed by the flow of traffic and by various traffic laws. One is tied down by various obligations: licence requirements, driver test, renewing registration, insurance, maintenance required for safety, monthly payments on purchase price, etc. Moreover, the use of motorised transport is no longer optional. Since the introduction of motorised transport, the arrangement of our cities has changed in such a way that the majority of people no longer live within walking distance of their place of employment, shopping areas and recreational opportunities, so that they have to depend on the automobile for transportation. Or else they must use public transportation, in which case they have even less control over their own movement than when driving a car. Even the walker's freedom is now greatly restricted. In the city he continually has to stop and wait for traffic lights that are designed mainly to serve auto traffic. In the country, motor traffic makes it dangerous and unpleasant to walk along the highway. (Note the important point we have illustrated with the case of motorised transport: When a new item of technology is introduced as an option that an individual can accept or not as he chooses, it does not necessarily remain optional. In many cases the new technology changes society in such a way that people eventually find themselves forced to use it.)

Source: www.thecourier.com/manifest.htm, 9 November 2005

cially when citizens engage in the debates instead of leaving these issues to experts.

The reasoning of many technological optimists is rather similar to that of Ellul and the Unabomber as they often paint a picture of a high-tech future as if this future were not human-made, but an inevitable reality.

A typical line of argument goes as follows:

- The pool of scientific knowledge is growing continuously

- The development of technology can utilise ever more scientific knowledge and other technologies

- This will lead to better technologies (which will in turn stimulate scientific growth and the development of other technologies)

- Citizens have to adapt to these new technologies

This kind of reasoning (whether optimistic or pessimistic) is called **technological determinism**: the state of technological development determines the nature of a society rather than vice versa. The technological deterministic view of the world is reflected in such historical characterisations as the 'Stone Age', the 'Bronze Age' or the 'age of the automobile'.

Technological determinism is questionable. It assumes that the relationship between science, technology and society is only unidirectional. As the example of the steam engine shows, technology results not merely from the simple application of scientific principles. Various analyses show, moreover, that the creation of new technologies is not a process with a single ('optimum') outcome, as Ellul holds, but is the result of a process of choice—a characteristic Ellul claimed only for traditional technology.[23] However, technological determinism cannot be rejected completely.

Social constructivism

Technological determinism is completely rejected in the social construction of technology (SCOT) model. In this model, technologies are considered to be social constructions, i.e. technologies have been given shape by the demands of various social groups. Central to this view is the notion that people can attribute different meanings to the same artefact (human-made object). The car, for instance, might be a means of transport for its owner, but it is also an object to demonstrate status or lifestyle. For pedestrians, the car may mean danger and, for administrators, it may mean an object for taxation.

Various groups may therefore try to influence the development of the technology in different directions. These might sometimes coincide but might also lead to controversy.

Technological artefacts possess certain stability, as they are building on past experiences and existing knowledge. Changes in technology will therefore not coincide with sometimes rapidly changing demands in society, but rather lag behind. The reverse also occurs: society adapts itself to the possi-

---

23　There are other forms of determinism that reduce processes of change to one single factor. One might argue for instance that Marxism contains a form of socioeconomic determinism reducing history to class conflict: K. Marx, 'The History of All Hitherto Existing Society is the History of Class Struggle', *Communist Manifesto* (1848).

bilities that new technologies offer, but these adaptations may take decades (e.g. the use of computers by older people).

These delays create a dynamic pattern of technology–society interaction. As many technologies fail and certain demands of society cannot be fulfilled, a co-evolutionary pattern of development emerges.

For the constructivist, the stability of technological artefacts is explained by the stability of the social preferences in society. The SCOT model is explained further in Chapter 9.

The debate between 'constructivism' and 'determinism' is not just of scholarly interest. For technological determinists, attempts to influence science and technology for commercial, ethical or political reasons are futile, producing no more than minor ripples in the pond. For social constructivists, society is constantly deciding on the future shape of technologies (even people that claim to be technological determinists). For social constructivists, shaping technologies for sustainable development is a realistic track to pursue.

Technological determinism and social constructivism as recurring stages

Between these two rather extreme positions, concepts have been developed that aim to integrate the determinist's main assertion (i.e. technology is self-propelling) with that of the constructivist (i.e. technology is the outcome of socially determined choice). These new conceptions seek to account for the conditions under which:

- Technological change is propelled predominantly by social forces

- Social forces can scarcely influence that process

Stability of technologies can be explained by the various different environments in which technologies are embedded:

- **Socio-economic environment**: a technology must meet the demands imposed on it by all the relevant social actors in its environment (for a car: price, status, comfort, safety, appeal, speed, etc.). New technologies must, in other words, solve the problems that actors think can be solved by the artefact in question

- **Physical environment**: every artefact is adapted to other artefacts, technological infrastructure, maintenance systems, energy sources, etc. For example, cars must fit on roadways, use available, standardised fuels, be fitted with familiar steering mechanisms and meet various performance standards before they are approved. New technologies must be compatible with these existing conditions

- **Technological knowledge base**: technologies are based on existing know-how, rules and accepted paths for further innovation that are accepted within a particular technological community. New technological artefacts arise from the state of knowledge and the shared beliefs regarding possible improvements within the community of practitioners designing and constructing them. More radical technological change therefore implies changing mainstream ideas within technological communities or breaking their power by creating an alternative technological community to take its place

Since major upheavals in all three realms rarely occur simultaneously, radical innovations are likewise rare. For example, an alternative for the car should be socio-economically viable (in terms of costs/performance ratio) and adapted to the existing infrastructure (roads, fuel, regulation, etc.). We should also have the ability to design it.

The various forces favouring technological stability mean that technological artefacts generally change only incrementally and over long periods of time. As in biological evolution, however, a technology may become extinct or split into several species adapted to specific niches and circumstances.

In the large-scale technological systems of today, social institutions and technological hardware form a seamless web and any distinction between the 'social' and 'technological' dimensions of these systems becomes futile. Particularly when systems fail, attempts are made to blame casualties on either 'human' or 'technological' factors. Such attempts are doomed to failure, though, for it is in fact impossible to distinguish the human and technological factors in any given technological system. Is it the hardware that is not properly adapted to the humans operating, administering or maintaining it, or are the humans not functioning in accordance with the demands set by the hardware they are dealing with?

Technological changes are sometimes slow and can easily lag behind the rapid pace of change in society as a whole. However, the creativity of technologists also leads to new products and systems that revolutionise social life such as mobile phones and computers. This is not to say that every new revolutionary technology is accepted by society. Indeed, many new technologies were not accepted and are hardly remembered. Civil aircraft with vertical take-off and landing, or soluble tablets to replace toothpaste are just two of the vast array of technologies that have been rejected. In the case of nuclear power, the issue of acceptance has still not been settled.

It might therefore be argued that, in so far as technology can be distinguished from its social environment, the relationship between them is a (co-)evolutionary one—they adapt to one another, but there are many mismatches. Such mismatches occur especially in times of rapid change, due either to massive breakthrough of new technologies, as in the case of information technology (IT) in the 1980s, or to rapid changes in society's preferences.

The technologies applied to solve environmental problems are products of:

- The present and the past

- Society's norms and values

- Technologists' experiences and paradigms

They shape our common future while reflecting contemporary standards and interests. To develop the technologies that humankind needs for sustainable development is therefore not only a formidable challenge for technologists, but for our entire technology-dependent civilisation.

Technology ideology and responsibility

Political choice and controversy

Decision-making with respect to technologies is probably more important for the long-term future of society than any other political issue today. Although technologies are not very often in the media focus (apart from major accidents), the decisions that are made in their design, development and introduction are crucial for the future of society.

Decisions on technology are not neutral. Technology is a political issue and so are sustainable technologies. For example: one might seek the solution of the environmental problems created by current agricultural practice either in organic farming (requiring large areas) or in extensive high-tech farming such as so-called pig factories (being highly efficient, with low emissions, but removing animals from their natural environment).

However, new technologies are very often not recognised as political issues requiring due debate.

Striking examples of political technologies are the extraordinarily low overpasses over the Parkways on Long Island, New York, which have as little as nine feet clearance.[24] They were deliberately designed and built that way by Robert Moses, the master builder of New York from the 1920s to the 1970s. He built his overpasses to discourage the presence of buses on his parkways. These reasons reflect his social class bias and racial prejudice. Car-owning whites of the 'upper' and 'comfortable middle' classes, as he called them, would be free to use the parkways for recreation and commuting. Poor people and blacks, who normally used public transit, were kept off the roads because the 12-foot-tall buses could not handle the overpasses. One consequence was that racial minorities and low-income groups had limited access to Jones Beach. Moses made doubly sure of this result by vetoing a proposed extension of the Long Island Railroad to Jones Beach.

---

24 L. Winner, *The Whale and the Reactor: A Search for the Limits in an Age of High Technology* (Chicago: University of Chicago Press, 1986). This example is derived from: R.A. Caro, *The Power Broker: Robert Moses and the Fall of New York* (New York: Vintage, 1975).

If the political nature of technological decisions becomes evident, technology becomes the focal point of debate. But, because the political nature of technological decisions is often hidden, this is often after official decisions have been made. Public controversies are a nuisance to decision-makers, but they are highly important for the democratic quality of the decision-making process. Controversies stimulate the scrutiny of the arguments and hence improve the clarity of the subject among the public.[25]

In these controversies, the **control dilemma** is often important. According to Collingridge, decision-makers on technologies are caught in a 'control dilemma':

> attempting to control a technology is difficult, and not rarely impossible, because during its early stages, when it can be controlled, not enough can be known about its harmful social consequences to warrant controlling its development; but by the time these consequences are apparent, control has become costly and slow.[26]

One of two main ways to solve this problem is by attempting, in the design phase, to gain insight into (side) effects occurring later on. This is, according to Collingridge, a dead end. Historical examples show that the attempt to predict all side effects is doomed to fail. Collingridge thus proposes an alternative way, which gears itself towards the design phase and attempts to integrate the idea that decisions have to be made based on (partial) ignorance. Hence technologies should be flexible. But what should they be flexible for? Very often the norms and values for evaluation of the effects are also unclear and could shift over time.

Public controversies can contribute to a better understanding of a technology and its effects. Various reports covering all kinds of technological details are written during such events, but these reports rarely aim to discuss or exchange arguments. The contestants generally try to discredit each other by pointing at unsound expertise and to prove the interests or ideological basis of their opponents.[27] In practice, such controversies are never decided by techno-scientific arguments.

However, a sound techno-scientific argumentation is a precondition for public and political credibility. Technological controversies are sometimes decided by politicians but, as politicians are often made to feel insecure by keen debate in society as well as among experts, controversies often either

25 See for example: A. Mazur, *The Dynamics of Technical Controversy* (Washington, DC: Communications Press, 1985) and A. Rip, 'Controversies as Informal Technology Assessment', *Knowledge: Creation, Diffusion, Utilisation* 8 (1986): 350.

26 D. Collingridge, *The Social Control of Technology* (London: Pinter Publishing, 1980).

27 Such an analysis regarding the role of physicists in the nuclear energy controversy in Austria was made in: H. Nowotny, *Kernenergie: Gefahr oder Notwendigkeit? Anatomie eines Konflikts* (Frankfurt am Main, Germany: Suhrkamp, 1979).

end with external developments such as the nuclear accident at Chernobyl or they gradually disappear.[28]

The responsibility of engineers and scientists

Scientist and engineers have considerable responsibility in public decision-making on techno-science. They are often the first to recognise the dangers to the public of new techno-scientific developments. For example, the discussion on the dangers of DNA research was started by scientists. Robert Pollack, a US virologist was the first to voice concerns to his colleagues; this led to a letter of warning published in *Science* in 1974.[29]

Moreover, as expertise gives credibility to viewpoints in the public debate, it is important that underprivileged groups such as non-governmental organisations (NGOs) and citizens' initiatives also have access to scientific expertise. 'Science shops' can play an important role in this. These small units (generally within universities) can offer the scientific expertise needed by NGOs and citizens' groups.[30]

The public responsibility of scientists and engineers can sometimes create moral dilemmas. The interests of employers, private industry or public organisations alike may sometimes be at stake if a scientist or engineer speaks up in the public interest. These people often run into trouble, endangering their career or even their job. Codes of conduct published by professional associations are important to help people reflect on dilemmas (not primarily to prescribe a specific conduct) and to support individuals that speak up.[31] Figure 5.10 presents the code prepared by the Institute of Electrical and Electronics Engineers (IEEE).

Science and technology in the daily life of rich nations

In day-to-day life, citizens in rich countries have become completely dependent on technologies. However, they are hardly aware of it and this is how it should be. In general, if they become aware of technological dependence, something is wrong. For example, the non-functioning of an Internet connection is a minor nuisance. A major problem is a power cut. At home, people manage eventually with the help of the social network but, when a

28 D. Nelkin, *Controversy: Politics of Technical Decisions* (Beverly Hills, CA: Sage Publications, 1984).
29 P. Berg, D. Baltimore, H. Boyer, S. Cohen, R. Davis, D. Hogness, R. Roblin, J. Watson, S. Weissman and N. Zinder, 'Potential Biohazards of Recombinant DNA', *Science* 185 (1974): 303.
30 Cf. the European Science Shops Network; www.scienceshops.org, 9 November 2005.
31 Cf. the Online Ethics Centre for Engineering and Science; onlineethics.org/codes, 9 November 2005.

IEEE (Institute of Electrical and Electronics Engineers) Code of Ethics

We, the members of the IEEE, in recognition of the importance of our technologies in affecting the quality of life throughout the world, and in accepting a personal obligation to our profession, its members and the communities we serve, do hereby commit ourselves to the highest ethical and professional conduct and agree:

1. to accept responsibility in making engineering decisions consistent with the safety, health and welfare of the public, and to disclose promptly factors that might endanger the public or the environment;

2. to avoid real or perceived conflicts of interest whenever possible, and to disclose them to affected parties when they do exist;

3. to be honest and realistic in stating claims or estimates based on available data;

4. to reject bribery in all its forms;

5. to improve the understanding of technology, its appropriate application, and potential consequences;

6. to maintain and improve our technical competence and to undertake technological tasks for others only if qualified by training or experience, or after full disclosure of pertinent limitations;

7. to seek, accept, and offer honest criticism of technical work, to acknowledge and correct errors, and to credit properly the contributions of others;

8. to treat fairly all persons regardless of such factors as race, religion, gender, disability, age, or national origin;

9. to avoid injuring others, their property, reputation, or employment by false or malicious action;

10. to assist colleagues and co-workers in their professional development and to support them in following this code of ethics.

Approved by the IEEE Board of Directors, August 1990

FIGURE 5.10 **IEEE code of ethics**

Source: www.ieee.org, 9 November 2005

blackout hits a whole city, the social organisation teeters on the brink of a total collapse. Such an event happened in New York in 1977:

> on the evening of July 13, 1977 lightning knocked out a major electrical transmission line [. . .] Within minutes of the electrical shutdown, looting broke out in widespread parts of the city, including the Upper West Side, East Harlem, and downtown Brooklyn. Police cars careened through dark streets, scattering crowds helping themselves to clothing, groceries, and furniture. In the South Bronx and Bushwick, fires burned out of control. By the time Consolidated Edison restored power the following evening, looters, rioters, and arsonists had caused an estimated three hundred million dollars in damages and the police had arrested more than three thousand people.[32]

Technology has intruded on everybody's life and most people seem perfectly happy about this. As we compare the lives of people in the rich world with those of people 500 years ago, technology makes them much more comfortable. People live longer and in a better condition, and there are far more of them. However, modern citizens still have concerns.

For example, we seem to accept the death toll from traffic accidents as a way of life. In the European Union alone, 42,500 people are killed every year and 3.5 million people are injured in traffic accidents.[33] After being a major political issue around 1970, the number of victims of traffic accidents fell considerable—partly through the introduction of safer technologies and partly by adapting the traffic system. Although traffic accidents are still an issue of serious concern, the victims are more or less considered to be the price that society accepts for its mobility.

Technologies are not just artefacts. In order to be able to use a technology, one has to know how to use the artefact, i.e. when to apply it, the proper way to activate it and the results it produces. We generally learn these things pretty fast, and are hardly aware that we learned them. But if this learning takes too much time or becomes too complicated, some people fail to achieve it (e.g. older people having trouble operating their computer or programming their video-recorder). This learning involves very minor issues, which become apparent only when somebody makes a mistake.

> When the railways introduced new doors in trains that closed automatically, sometimes people could be seen struggling with doors that did not close manually.

---

32   J.B. Freeman, *Working Class New York: Life and Labour Since World War II* (New York: New Press, 2000). See also: blackout.gmu.edu/events/tl1977.html, 9 November 2005.
33   European Transportation Safety Council (ETSC), *Intelligent Transportation Systems and Road Safety* (Brussels: ETSC, 1999).

In Barcelona, you can regularly see tourists struggling with the access gates to the metro, as they did not learn to feed the ticket into a slot at the left side of the gate instead of the right side.

Technological artefacts prescribe certain behaviour to us—otherwise they are pointless. These hidden prescriptions are called **scripts**. The clearest example of what happens if we do not know the 'scripts' can be seen in the 1970s comedy *Catweazle*.

Catweazle was an 11th-century magician who ended up in our world by mistake. What made us laugh in the comedy was that Catweazle attributed his own scripts to the artefacts that he found. A welding shield was interpreted as a magic helmet, and so the script that Catweazle used was to wear it screaming various spells.[34]

There was another interesting feature of Catweazle. He interpreted the functions of modern artefacts within his magic framework. Thus, the telephone became a magic speaking bone. In the traditional magic framework, every object was besouled.

Modern citizens completely lost this framework of interpreting artefacts. Telephones are techno-scientific instruments that operate according to scientific laws, though few would be able to explain those laws. A magic or divine interpretation of the forces of nature is hardly attractive to the modern citizen. For example, storms are phenomena that can be foreseen and explained by meteorology. Lightning strikes are an electrical discharge of clouds, which can be explained by electromagnetic laws. Science removed the magic from the daily life. There are no longer divine blessings and punishments; there are just the laws of nature and bad luck.

The modern citizen is sceptical of stories of miracles such as reports of sacred wooden icons shedding a tear or the search for the monster supposed to lurk in Loch Ness in Scotland. Unidentified flying objects (UFOs) seem to be miracle stories that took the shape of our techno-scientific world-view. However, the recurrence of these stories seems to point towards new problems and feelings of unease resulting from the very success of science and technology:

- Who are we as humans? Is our life determined merely by the genes that combined in our conception and the strokes of good and bad luck afterwards? And, if so, what does it make us?

- Scientists in Edinburgh, Scotland, succeeded in cloning a mammal, Dolly the sheep, who was born on 24 February 1997.[35] In prin-

34 For some Catweazle clips, see: www.propaganda.com.au/catweazle/downloads. shtml, 25 January 2005.
35 www.sciencemuseum.org.uk/antenna/dolly/index.asp, 10 March 2006.

ciple, humans can be cloned too. Since 1978, babies can be conceived outside the mother's womb. Eggs can be fertilised in a test-tube and the fertilised eggs can be genetically screened. Where does this lead us? We are increasingly able to select our own descendants. Do we want only perfect babies? What does this imply for people that are less than perfect?

● In May 1997, IBM's Deep Blue Supercomputer played a fascinating game of chess with the reigning world champion Garry Kasparov, winning the sixth deciding match.[36] What does this do for us? Are humans just a stage in the evolution that will be surpassed by more intelligent designs?

It is beyond the scope of this book to answer these questions—if indeed they can be answered at all. What is clear is that modern science and technology affect us so deeply that fundamental questions arise for every human being.

Questions, discussion and exercises

1. The well-known Moore's Law (Figure 5.11) predicts that the number of transistors per square inch on integrated circuit boards will double every two years.
   a. Is this a proof of technological determinism, i.e. that this technology progresses regardless of changing social factors or could you also interpret it as an result of social construction?
   b. Search the internet for statements that support these two interpretations. Identify who is making these statements.

2. Check the history of your own university or institute.[37]
   a. Who founded it?
   b. What role was it intended to play in society? In what regard were its educational programmes new at the time of its founders?

36 See www.research.ibm.com/deepblue, 9 November 2005.
37 If not available, search the websites of one of the older engineering universities such as: Massachusetts Institute of Technology (MIT), libraries.mit.edu/archives/mithistory/index.html, 9 November 2005; Imperial College, London, www3.imperial.ac.uk/portal/page?_pageid=73,369046&_dad=portallive&_schema=PORTALLIVE, 17 March 2006; TU Delft, the Netherlands, www.tudelft.nl/index.cfm/site/Organisatie/pageid/66142E6B-8C71-EDEB-0D57799545A36F7D/index.cfm, 9 November 2005; RWTH Aachen, www.rwth-aachen.de/zentral/english_history.html, 9 November 2005; ETH Zurich, www.150jahre.ethz.ch/program/ethistory/index_EN, 9 November 2005.

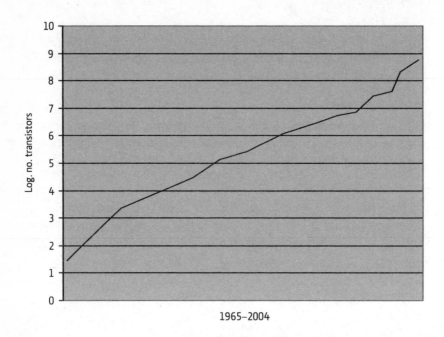

FIGURE 5.11 Moore's Law (transistors per integrated circuit)

Source: ftp://download.intel.com/pressroom/kits/events/moores_law_40th/MLTimeline.pdf, 9 November 2005

3. What is the control dilemma?

   a. Could you escape the control dilemma by making technologies flexible in regard to future changes or new knowledge?

   b. What would be the consequences of such a strategy for the development of new technologies?

   c. Discuss this with respect to the introduction of genetically modified crops or the development of nanoscale devices that could operate entirely on their own.

6 Measuring sustainability

This chapter outlines how sustainability can be measured. The complexities involved in measuring changes in social, natural and economic capital are enormous and there is no unique standardised method for measuring sustainability. Various measuring methods have been proposed. However, it is possible to detect some common features, which all have their origin in the UN Conference on Environment and Development held in Rio de Janeiro in 1992.[1]

Perspectives on sustainability differ in regard to the relationship between human development and nature. To measure sustainability, it is necessary to integrate spheres that have traditionally been measured separately.

The main objective of measuring sustainability is frequently to monitor the evolution of variables and indicators over time. In this way, we can understand where we are and where we are going. It helps us to choose targets for the future and to determine how far we are from where we want to arrive.

Comparison between countries or regions is also important. Indicators and indexes help to identify problem areas and to establish the path to find solutions.

To measure sustainability it is necessary to have:

- An appropriate methodological framework
- Contrasting and trustworthy data
- A strategy of appropriate communication
- A system of permanent evaluation

1   This conference is also known as the Earth Summit.

General principles

Sustainable development is not something that is achieved in a passive way. Decisions must be taken and strategies defined in order to establish a certain route for development. It is necessary to adopt a specific course of action, which has to be communicated to a range of actors. Those such as regional authorities or city councils should be able to monitor their problems and their progress towards solutions in order to make decisions.

It is important that decision-making is based on accurate information. To achieve this, it is necessary to create an information system that integrates the principles and approaches of sustainable development. The information that is generated by these systems must be easily accessible in order to facilitate public participation.

Due to the complexity of sustainable development, such information systems can never provide a complete overview of all its relevant aspects. Moreover, too much information can create an 'information overflow' for decision-makers. To reduce this complexity, indicators are needed that give an overview of specific segments of reality.

Strategies for measuring sustainability should be able to:

- Monitor the dynamics of problems

- Evaluate the consequences of proposed solutions

In this way, information systems can form the basis for a more appropriate method of taking decisions.

Indicators and indexes

Words such as index and indicator are often used interchangeably. This gives rise to interpretation errors. We define a **variable** as those characteristics of a system that can be measured.

Indicators represent a specific phenomenon that cannot be measured directly. They are obtained by the combination of different variables (e.g. energy intensity as energy consumption per capita, etc.). They should generally give enough information for a subjective evaluation of the problem. This evaluation is usually made by comparing the value of the indicator with a threshold value.

> Indicators are constituted to facilitate the process of decision-making. They help to measure and to gauge the progress toward the goals of sustainable development. On the other hand, they can be constituted as warning signals that prevent social, environmental and economic damages.[2]

2 United Nations, *Indicators of Sustainable Development: Framework and Methodologies* (9th session of the Commission on Sustainable Development held in New York, 16–27 April 2001; New York: United Nations Publications, 2nd edn, 2001); www.un.org/esa/sustdev/natlinfo/indicators/indisd/english/english.htm, 10 November 2005.

Indexes are values that communicate overall information on a specific problem. They are obtained by aggregating various indicators or variables that are assumed as components of the phenomenon under study. The indicators and/or variables are often weighted to create the index.

A life-cycle assessment (LCA; see Chapter 8) is an example of a product-related index.

Sustainability indicators

Sustainability indicators are tools that communicate information of a complex phenomenon such as gross domestic product (GDP), emissions of carbon dioxide (CO_2) per person, etc. They can be calculated per period and/or per area and allow one to evaluate the progress of a region/city/country towards a specific goal.

To allow policy assessments, it must be clear which value of the indicator is more or less desirable.[3] This requires a gradient that can have different forms:

- **Nominal scales** consist of only two values such as yes/no. Nominal scales provide little meaningful information, but are easy to agree on in case of controversial themes. For example, whereas the effectiveness of a national sustainability council may be questionable, their existence is easy to report

- **Ordinal scales** are based on a hierarchy of qualitative states, e.g. the quality of training of personnel, the transparency of decision-making processes or the possibilities for public participation in them. To apply these scales properly, the hierarchy has to be made explicit and the relative distances between the different classes defined. However, these distances are often based on value judgments and not easily agreed on

- **Cardinal scales** give quantitative information. If sustainable development goals are linked to a quantitative target, the distance towards this goal can be measured. Such indicators are called 'performance indicators'. To derive the scales, quantified targets have to be agreed on

Cardinal performance indicators are preferred, with ordinal indicators providing an alternative. In general, indicators have to be:

- **General**, i.e. not dependent on a specific situation, culture or economic organisation

---

3   Sustainable Europe Research Institute (SERI); www.seri.de, 10 November 2005.

- **Indicative**, i.e. truly representative of the phenomenon intended to be characterised

- **Sensitive**, i.e. they have to respond early and clearly to changes in what they are monitoring

- **Robust**, i.e. with no significant changes in case of minor changes in the methodology or improvements in the database

Sustainable development indicators need a benchmark (sustainable value) that gives an objective. The assessment process can then begin, in which sustainable and unsustainable tendencies are measured.

The process of information collection for decision-making is summarised in Figure 6.1.

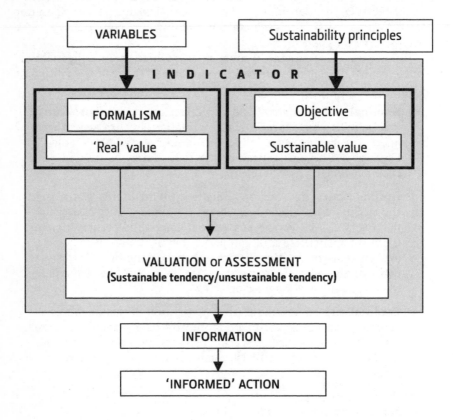

FIGURE 6.1 Information collection for decision-making

The Bellagio Principles

In November 1996, the International Institute for Sustainable Development (IISD) developed a set of principles to measure sustainable development known as the Bellagio Principles (Figure 6.2). These aim to assess progress towards sustainable development.

Methodological frameworks

Sustainable development indicators are generally presented in structured tables. The purpose is to articulate a coherent framework that facilitates the visualisation of the information. The use of frameworks is indispensable for a correct application of a group of indicators.[4]

There is currently no one universal framework that offers concrete rules regarding design, selection or development of indicators. Various factors (e.g. developed or developing country) determine the visualisation of reality and the scales of application of the indicators (local, regional, national, global, etc.). Major frameworks include:[5]

- Integrated national environmental–economic accounting systems or other national accounts-based recordings of stocks and flows of natural resources and environmental services—in most cases expressed in monetary terms

- Environmental statistics that list statistical variables in a systematic manner without establishing functional relationships among those variables

- Ad hoc environmental indicators using selected 'themes', 'issues' or 'subsystems' of models

- Overall policy frameworks or conventions, which were not designated for data collection but reflect the concerns to be monitored (e.g. Agenda 21)

4   United Nations Environment Programme (UNEP) and Department for Policy Coordination and Sustainable Development (DPCSD), 'The Role of Indicators in Decision-Making', joint paper by UNEP and the UN Division for Sustainable Development, DPCSD, presented at an International Workshop on Indicators of Sustainable Development for Decision-Making held 9–11 January in Ghent, Belgium (The Ghent Report; Geneva: UNEP, 1995): 6.
5   See G.C. Gallopin, 'Indicators and their Use: Information for Decision-making', in B. Moldan and S. Billhartz (eds.), *Sustainability Indicators: Report on the Project on Indicators of Sustainable Development* (Scientific Committee on Problems of the Environment (SCOPE) Report 58; Chichester: John Wiley, 1997): 13-27; www.icsu-scope.org/downloadpubs/scope58/cho1-introd.html, 10 November 2005.

1. Guiding vision and goals

Assessment of progress toward sustainable development should be guided by a clear vision of sustainable development and goals that define that vision.

2. Holistic perspective

Assessment of progress toward sustainable development should:

- Include review of the whole system as well as its parts
- Consider the well-being of social, ecological, and economic sub-systems, their state as well as the direction and rate of change of that state, of their component parts, and the interaction between parts
- Consider both positive and negative consequences of human activity, in a way that reflects the costs and benefits for human and ecological systems, in monetary and non-monetary terms

3. Essential elements

Assessment of progress toward sustainable development should:

- Consider equity and disparity within the current population and between present and future generations, dealing with such concerns as resource use, over-consumption and poverty, human rights, and access to services, as appropriate
- Consider the ecological conditions on which life depends
- Consider economic development and other, non-market activities that contribute to human/social well-being

4. Adequate scope

Assessment of progress toward sustainable development should:

- Adopt a time horizon long enough to capture both human and ecosystem timescales, thus responding to needs of future generations as well as those current to short-term decision-making
- Define the space of study large enough to include not only local but also long-distance impacts on people and ecosystems
- Build on historic and current conditions to anticipate future conditions—where we want to go, where we could go

5. Practical focus

Assessment of progress toward sustainable development should be based on:

- An explicit set of categories or an organising framework that links vision and goals to indicators and assessment criteria

FIGURE 6.2 **The Bellagio Principles** (continued opposite)

Source: www.iisd.org/measure/principles/bp_full.asp, 10 November 2005

- A limited number of key issues for analysis
- A limited number of indicators or indicator combinations to provide a clearer signal of progress
- Standardising measurement wherever possible to permit comparison
- Comparing indicator values to targets, reference values, ranges, thresholds, or direction of trends, as appropriate

6. Openness

Assessment of progress toward sustainable development should:

- Make the methods and data that are used accessible to all
- Make explicit all judgements, assumptions and uncertainties in data and interpretations

7. Effective communication

Assessment of progress toward sustainable development should:

- Be designed to address the needs of the audience and set of users
- Draw from indicators and other tools that are stimulating and serve to engage decision-makers
- Aim, from the outset, for simplicity in structure and use of clear and plain language

8. Broad participation

Assessment of progress toward sustainable development should:

- Obtain broad representation of key grass-roots, professional, technical and social groups, including youth, women and indigenous people—to ensure recognition of diverse and changing values
- Ensure the participation of decision-makers to secure a firm link to adopted policies and resulting action

9. Ongoing assessment

Assessment of progress toward sustainable development should:

- Develop a capacity for repeated measurement to determine trends
- Be iterative, adaptive, and responsive to change and uncertainty because systems are complex and change frequently
- Adjust goals, frameworks, and indicators as new insights are gained
- Promote development of collective learning and feedback to decision-making

10. Institutional capacity

Continuity of assessing progress toward sustainable development should be assured by:

- Clearly assigning responsibility and providing ongoing support in the decision-making process
- Providing institutional capacity for data collection, maintenance and documentation
- Supporting development of local assessment capacity

The main framework models for the integration, design and presentation of indicators are described below.

PSR and DPSIR models

The Pressure–State–Response (PSR) framework model is used by international institutions—especially the Organisation for Economic Co-operation and Development (OECD).[6] The indicators are structured according to three basic categories (Figure 6.3).

- **Pressure** refers to human activity that creates some type of pressure on natural systems such as emissions of greenhouse gases, production of waste, etc.

- **State** refers to the changes in the quality and in the quantity of natural resources, and the measure of these changes evaluated in a certain period of time. This gives us the 'state' of the natural system. Examples of state indicators include global mean temperature, threatened species and concentrations of substances

- **Response** refers to the answer in terms of policies or specific actions that were taken as a response to changes detected in the natural system such as recycling rates, international commitments, etc.

As well as the OECD, the PSR model and its variants have been used by other institutions including the United Nations Commission on Sustainable Development (UNCSD),[7] Eurostat[8] and the European Environment Agency.[9]

However, the PSR model has a number of disadvantages:

- It is focused on environmental issues

- It tends towards a linear visualisation

- It is focused on processes of tension (decline of forests, climate change, etc.), which tends to lead to the development of fundamental options for remedy

- It tends to work back from the requested state without taking external developments into account

An enlarged version of this model is called DPSIR (Driving forces–Pressure–State–Impact–Response). It widens the PSR framework by adding the Driving forces of Pressure and Impacts of state on society (Figure 6.4).

6   www.oecd.org, 10 November 2005.
7   www.un.org/esa/sustdev/csd/csd.htm, 10 November 2005.
8   Statistical Office of the European Communities; europa.eu.int/comm/eurostat, 10 November 2005.
9   www.eea.eu.int, 10 November 2005.

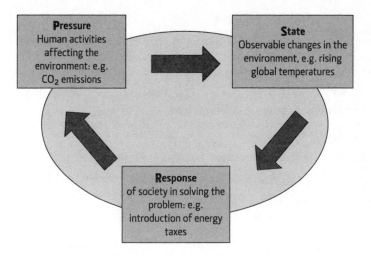

Figure 6.3 **PSR model**

Source: Eurostat

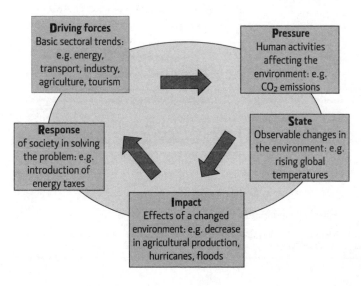

Figure 6.4 **DPSIR model**

Source: Eurostat

Hierarchy model based on principles and objectives

One possibility is to introduce a hierarchy that leads from a general objective to more specific principles and criteria to indicators. In this way, the resulting set of indicators can be divided into modules that refer to the same sustainability principle such as:

- Sustainable economic development
- Eco-efficiency
- Sustainable use of resources
- Precautionary principle

Figure 6.5 shows such a model.

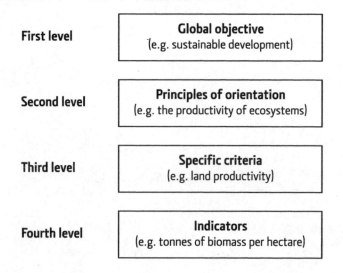

FIGURE 6.5 Hierarchical model

Source: modified from R. Quiroga, *Indicadores de Sostenibilidad Ambiental y Desarrollo Sostenible: Estado del arte y perspectivas* (Serie Manuales CEPAL; UN, 2001)

Framework of themes and sub-themes

This framework, which was drawn up by the UN World Commission on Sustainable Development (CSD), recognises four key dimensions of high priority:

- Environmental

- Social

- Economic

- Institutional

Each of these dimensions is divided into themes and sub-themes that reflect the main priorities established in the chapters of Agenda 21.[10] Finally, each sub-theme leads to one or more indicators (Figure 6.6). The set of indicators obtained presents the four dimensions, 15 themes and 38 sub-themes (Table 6.1).

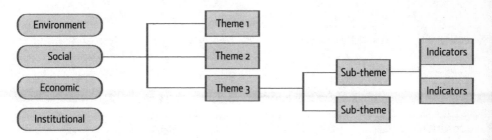

FIGURE 6.6 **CSD framework**

We are seeking indicators that reflect how well basic needs are satisfied under given circumstances. There are various ways of determining fundamental environmental properties.

In physical environments, we can analyse the physical signals we receive, e.g. using various instruments. Six fundamental properties of relevance are found (Figure 6.7).

As an example, let us examine these properties for a family:

- **Normal environmental state.** A family living in a small town in a European country has to deal with specific economic, social, cultural, legal and political environments that are different from those in, say, India

- **Resource scarcity.** The family needs money, water, food, electricity, consumer goods, medical services, sanitation, etc., all of which can be secured only with considerable effort

- **Variety.** The family has to exist in an environment containing a host of very different actors and factors, e.g. neighbours, friends,

10 Adopted at the Earth Summit in 1992; www.un.org/esa/sustdev/documents/ agenda21/index.htm, 10 November 2005.

SOCIAL DIMENSION

Themes	Sub-themes
Justice	• Poverty • Equity
Health	• Nutritional state • Mortality • Sanitation • Drinking water • Health benefits
Education	• Educational level • Illiteracy
Housing	• Living conditions
Security	• Crime
Population	• Population dynamics

ENVIRONMENTAL DIMENSION

Themes	Sub-themes
Atmosphere	• Climate change • Ozone layer • Air quality
Land	• Agriculture • Forests • Desertification • Urbanisation
Oceans and coasts	• Coastal areas • Fisheries
Freshwater	• Water quantity • Water quality
Biodiversity	• Ecosystems • Species

INSTITUTIONAL DIMENSION

Themes	Sub-themes
Institutional framework	• Strategies for sustainable development • International co-operation
Institutional capacity	• Access to information • Communications infrastructure • Science and technology • Preparation for, and aid capacity in natural disasters

ECONOMIC DIMENSION

Themes	Sub-themes
Economic structures	• Economic development • Trade • Finance
Patterns of consumption and production	• Energy use • Production and management of waste • Transport

TABLE 6.1 CSD themes and sub-themes

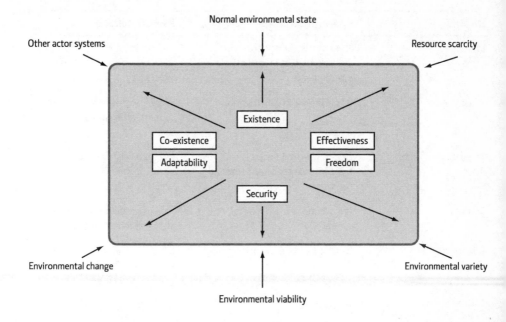

Normal environmental state

Other actor systems

Resource scarcity

Existence

Co-existence

Effectiveness

Adaptability

Freedom

Security

Environmental change

Environmental variety

Environmental viability

FIGURE 6.7 Fundamental properties of system environments and their basic orienting
counterparts in systems

bosses, colleagues, authorities, different shops and a multitude of
social and cultural offerings

- **Variability.** The variety also changes rapidly—a new neighbour
  moves in, members of the family become ill, change their friends,
  lose their jobs, or have to move

- **Change**. There is not only normal variability. Economic and social
  conditions can change and new technologies can become available
  in the house and the workplace

- **Other systems.** The family has to care for their pets and ageing parents and accommodate the interests of an employer, neighbours or
  other drivers in traffic

Measurement tools

How are aspects of sustainable development measured? LCA is especially tar-
geted at products and is therefore discussed in combination with product

Orientor	System performance	Possible indicators
Existence	Is the system compatible with and able to exist in its particular environment?	Availability of shelter, clothing, food, water, sanitation, life expectancy
Effectiveness	Is it effective and efficient?	Work hours necessary for life support, efficiency of resource use
Freedom of action	Does it have the necessary freedom to respond and react as needed?	Income level, job opportunities, health, mobility
Security	Is it secure, safe and stable?	Safe neighbourhood, savings, insurance, social security scheme
Adaptability	Can it adapt to new challenges?	Education and training, flexibility, cultural norms
Co-existence	Is it compatible with interacting sub-systems?	Social skills, compatibility of language and culture
Psychological needs	Is it compatible with psychological needs and culture?	Emotional stress, anxiety, dissatisfaction, family quarrels

TABLE 6.2 Finding indicators for the viability of a family

design in Chapter 8. In this section, we examine the ecological footprint method and visual representation methods.

Ecological footprint

One of the best-known sustainable development indicators is the ecological footprint. This method produces an index that gives a quantitative reference of the way of life of a (certain group of) person(s), a region or a country.[11]

This index is not measured in monetary units, but in surface area. The ecological footprint is the quantity of land that a person needs, directly or indirectly, to:

- Consume products and services

- Produce resources and assimilate waste

11 W. Rees and M. Wackernagel, *Our Ecological Footprint: Reducing Human Impact on the Earth* (Gabriola Islands, BC: New Society Publishers, 1996).

This in turn means the land needed to produce the food and the materials for housing, buildings, roads, infrastructure and the trees that regenerate the CO_2 produced by burning fossil fuels.

The name 'ecological footprint' therefore refers to the land area that people, a country, a region or a city would use if they were sustainable.

The ecological footprint is the inverse concept of the carrying capacity of a region. The carrying capacity measures the quantity of people that can live within the given area, whereas the ecological footprint measures the quantity of land that a given group of people needs to live.

The size of the ecological footprint depends on a number of factors, including:

- The development and lifestyle of the subject

- The quantity of utilised resources

- The utilised technology

- Social and economic organisation

It would be desirable that the ecological footprint of a country should not surpass its territory. In that case, the country could provide for itself in a sustainable way (including trade).

Regrettably, many countries—most of them industrialised—have a much larger ecological footprint than their own territory. They consume many imported resources. An extreme example is the Netherlands, which has an equivalent footprint of 15 times its territory. This is caused by the high population density combined with high oil consumption and major food imports such as soy fodder for pig farming. Japan's ecological footprint also greatly exceeds its territory.

In a symmetrical way, many developing countries have much smaller ecological footprints than their territory; they do not exploit all their territory and they export many of the resources that they produce.

Table 6.3 shows the ecological footprints of various countries. The mean world ecological footprint per person is 2.8 hectares. This is much more than the area available given the bio-capacity of the Earth. The area that we use is larger than future generations will have. In other words, the ecological footprint indicates that the current way of living is not sustainable.

However, using the ecological footprint as an indicator has some disadvantages:

- The calculation of the area of the ecological footprint is not accurate because it is impossible to make an exhaustive list of the various ways in which we use the Earth

- Not all social costs are included in the indicator. For example, replacing monuments in overcrowded cities such as Gaudi's famous cathedral, the Sagrada Familia, in Barcelona with trees would reduce the ecological footprint, but it is not desirable

Country	Ecological footprint per capita (ha)
USA	10.3
Australia	9.0
Canada	7.7
Russia	6.0
Sweden	5.9
Germany	5.3
Netherlands	5.3
Great Britain	5.2
Belgium	5.0
Japan	4.3
Italia	4.2
Austria	4.1
France	4.1
Spain	3.8
Portugal	3.8
Venezuela	3.8
Brazil	3.1
Thailand	2.8
México	2.6
Chile	2.5
Turkey	2.1
Colombia	2.0
Peru	1.6
Nigeria	1.5
China	1.2
Egypt	1.2
Ethiopia	0.8
India	0.8
Pakistan	0.8
Bangladesh	0.5
Mean world ecological footprint	2.8
Available bio-capacity	1.7

TABLE 6.3 World ecological footprint, 1997

Source: M. Wackernagel, N. Chambers and C. Simmons, *Sharing Nature's Interest: Ecological Footprints as an Indicator of Sustainability* (London: Earthscan, 2000)

- The ecological footprint is a 'reductionist' indicator. A single indicator (money or area) cannot describe everything. It is necessary to use multiple approaches

Visual tools: the Dashboard and the Compass

A number of visual tools have been developed to measure sustainability. Two of the most important are:

- Dashboard of Sustainability[12] developed by the IISD
- Compass Index of Sustainability[13] developed by AtKisson Group

Dashboard of Sustainability

The Dashboard of Sustainability is an online tool that visualises the performance for sustainable development and helps politicians or citizens to make decisions.[14] The software provides a user-friendly format that allows people to compare the sustainability performance of countries, regions and communities.

The Dashboard is based on the UNCSD methodological indicators framework, divided into 19 social, 20 environmental, 14 economic and 8 institutional indicators (Figure 6.8).

Compass Index of Sustainability

The Compass offered by the AtKisson Group (a US/European consultancy) is a software package for developing a measure of genuine progress at different levels, e.g. community, business or organisation. It encompasses four aspects of sustainability:

- Nature
- Economy
- Society
- Well-being

The methodology allows all stakeholders to participation in every Compass point. They are introduced to sustainability concepts and principles, develop a set of assets and concerns, and then develop the indicators that will reflect critical long-term trends in every relevant area. They also explore the linkages between indicators and prepare the ground for creating co-operative action programmes.

12   esl.jrc.it/envind/dashbrds.htm, 10 November 2005.
13   www.atkisson.com/accelerator/#compass, 10 November 2005.
14   www.iisd.org/cgsdi/intro_dashboard.htm, 10 November 2005.

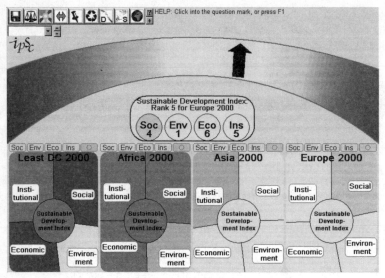

FIGURE 6.8
Dashboard of Sustainability

Source: esl.jrc.it, 10 November 2005   Special thanks to Jochen Jesinghaus

This tool provides a feedback to the stakeholders engaged in the decision-making process. This means that updates can be published on a regular basis and allows new sets of relationships among stakeholders to be developed, which might lead to new initiatives.

Technology indicators

Technology can also be measured for its contribution to sustainable development. The indicators for technology and sustainable development can be classified into two main categories:

A. Indicators of the relationship of a technology with the natural system

B. Indicators of the relationship of a technology with the social system

The indicators of the first group can be split into two groups of basic indicators:

A1. **Eco-efficiency indicators** that show the grade of dematerialisation of the technological system. This group includes indicators that refer to resource properties, e.g. if a resource is renewable or not

A2. **Indicators of environmental quality** that show the impact on the environment generated by the technology, e.g. emissions to atmosphere, water or land

The LCA of a product (see Chapter 8) is an indicator that performs this task. The indicators of the second group include those that provide information about the social effects of the technology:

B1. **Social aspects**, such as creation of employment by the technology

B2. The degree of **decentralisation** of the technology

B3. The degree of **complexity** of the technology and the degree of education required for its use

B4. The **investment** required for the technology

Indicators that measure the social effects of technology are not generally available. The social effects of a technology are usually dealt with by technology assessment (see Chapter 4). However, scorecards such as social compatibility analysis[15] do exist.

Checklists can be useful for ensuring that all relevant effects are retrieved and can be used to create an indicator. However, this does not happen very often because a single issue might make the technology unacceptable regardless of the score for the other issues. Figure 6.9 shows an example checklist.

Questions, discussion and exercises

1. Calculate your personal ecological footprint at www.ecofoot.org. Compare it with the footprint of your classmates. Develop options to reduce your footprint by 30%.

2. Apply the checklist for social sustainability effects of technology to an example of your choice from the following list:
   - Nuclear fusion
   - Large-scale near-shore wind turbine parks
   - High-speed rail transport
   - Hybrid cars
   - Unbreakable bicycle locks

15 Developed by V. Carabias-Hutter and H. Winistorfer, School of Engineering, Zurich University of Applied Sciences Winterthur (ZHW), Switzerland; zsa.zhwin. ch/cms/zsa/upload/pdf/SCA_poster.pdf, 10 November 2005.

A. How acceptable is the research and development work that accompanies the formation of the new product and its production process?

1. Do counteracting social forces exist against the methods used in the research and/or development work or against the collection and storage of certain data?

2. Is the research and development work scientifically interesting or does the development of this technology provide a special contribution to (a) technical and/or scientific discipline(s)?

3. Can it be foreseen that investing in the development of a technology at this moment could prevent a better alternative from being developed in the future?

B. How acceptable is the new product in itself?

1. Are any social values connected to the product in itself, or the product that it replaces?

2. Is the product considered unacceptable in the ethical system of specific religious or cultural factions?

3. Will the costs of the new product be likely to attract criticism?

4. What effects on the environment will the product have? Which changes in behaviour will the product cause, and what environmental effects result from that?

5. What are the risks the use of the product entails, both for the user and others?

6. Does the use of the product clash with habitual behavioural patterns of large groups of people?

7. Are there financial or psychological barriers that hinder acceptance of the product?

8. Does the product permit new (economic or otherwise) activities? How should these activities be judged?

9. Does the product threaten existing activities, which hold a certain social or cultural value?

10. Does the product influence the social structure (private life, local community, cultural region)?

11. Does the product have any other (possible) uses than the one primarily intended for it?

FIGURE 6.9 **Example of checklist for social sustainability effects of technology**
(continued opposite)

C. How acceptable is the production of the new product?

1. Are any ethical standards and/or values threatened in production?
2. Are the working conditions in the production process acceptable?
3. Which are the physical effects of the production facility on the environment?
4. Which are the expected (primary and secondary) effects of the production on employment? What level of schooling is required for personnel?
5. What other consequences does the product have for the local environment?
6. What are the social implications of the production for the local community?
7. Does a potent breeding ground for local activism exist?
8. Can choosing a suitable location drastically reduce negative effects?
9. Which (either existing or planned) economic activities are threatened by production?
10. Is the existing balance of power influenced by new production? Consider the following relations between:
    - Employees (or unions) and employers
    - Different producers
    - Producers, clients and suppliers
    - Government and industry sector
    - Different governmental institutions
11. What does new technology mean for the development of third-world countries? Are relations between global trade blocs influenced by production?

    – A worldwide data analysis and exchange system that could reduce corporate administrative fraud

3. Look through a report on the environmental performance of your city or town. Suggest three variables that are easy to measure and which could be used to create a sustainable development indicator for the city or town. Discuss the advantages and disadvantages of using these three variables.

4. Fill out the DPSIR for the problem of climate change. Identify areas where there are differences between countries.

7 Sustainable development and the company: why, what and how?

Companies exist to make a profit. But is the drive for profit the only motive that determines corporate action?

In small companies, owners have to earn their living from the company. In larger enterprises, the shareholders require money in return for their investment. But if we depict the company solely as a machine designed to squeeze the maximum amount of money out of the pockets of citizens or other companies, why should society tolerate them?

Do we need detailed laws to protect us from excessive greedy companies? Do we need a large organisation to police companies? Incidences of corporate fraud suggest that sometimes we do. But more regulation stifles economic activity and can never rule out every form of misconduct. Therefore, we need another course.

This chapter analyses how corporations deal with their responsibilities and what they can contribute to sustainable development through various aspects of their business.

Sustainability: why should companies care?

Companies are part of society and society is part of them; all employees also have private lives and participate in community activities. Without prosperous communities, there are no prosperous companies. So the least a company should do is respect the culture, norms and values of the communities of which it is part.

A company should understand that its income is based on the value it creates for society. A successful company produces value more efficiently than its competitors and therefore makes a larger profit. But society determines this value through laws, taxes and the market. Hence, no company can afford to live at war with society.

But there is more. Assuming a company is governed strictly by the principle of financial optimisation also assumes that this can be done unambiguously. But this is not the case:

> If you have an apple tree, you might cut it for the wood, rent it to somebody through an annual auction, or pick apples yourself to sell on the market. The first activity will give you the quickest return. The last activity will probably yield the highest return in the long run as the harvester will probably take care that the tree will remain in good shape and no auctions have to be organised.

A wise company will opt for optimised returns in the long run. But, as the famous economist John Maynard Keynes said, 'in the long run we are all dead',[1] and thus short-term policies will always be tempting. Moreover, any long-term policy will fail if the company is not able to survive short-term business conditions.

> Shifting from the worries of everyday profit-making to running a sustainable company that derives a profit from long-term value creation for society requires a change in perspective. It does not require one to abandon entrepreneurial activity; on the contrary, it is an extra opportunity and challenge for the entrepreneur. That's what sustainable entrepreneurship is all about.

A company's societal and economical responsibilities are more than just making a profit. The role of companies is to create value by using scarce goods (capital, labour, knowledge, management quality and natural resources) in an effective and efficient manner to produce goods and services that fulfil needs and contribute to social well-being. The employment created is an important means of providing people with an income, together

---

1 After claiming the long-term advantages of anti-cyclic economic policies, he used these words to admit the temptations posed by short-term policies.

with social and personal development. The principle of 'corporate citizenship' summarises this idea nicely:

> A company should behave like a good citizen in business. The law does not (and cannot) contain or prescribe the whole duty of a citizen. A good citizen takes account of the interests of others and tries to exercise an informed and imaginative ethical judgement in deciding what he should and should not do. This, it is suggested, is how companies should seek to behave.[2]

Sustainable entrepreneurship is a worldwide trend among companies. Various companies aim to achieve it and a number of organisations allow companies to exchange their experiences. The main one is the World Business Council for Sustainable Development (WBCSD)—a coalition of 170 international companies united by a shared commitment to sustainable development.[3]

Although sustainable entrepreneurship is interpreted in different ways, this is not a problem. The enormous diversity of business conditions and the unique set of actors and factors that a company deals with will result in different actions. But sustainable entrepreneurship generally implies that corporations are willing and able to behave responsibly towards society.

Sustainable entrepreneurship is not something that came about instantaneously. A long history of employers has sought voluntarily to contribute to society.

Improving the conditions of workers was important for some 19th-century entrepreneurs. For example, Robert Owen was the manager and part-owner of New Lanark Cotton Mills near Glasgow at the beginning of the 19th century. He built houses for his workers and provided education for the community. Children were not allowed to work before the age of ten. The workers were paid during illness and at old age. All corporal punishment was forbidden.[4]

In France, F.M.C. Fourier (1772–1837) came up with the idea of establishing productive communities in which people were compensated according to their contribution to the community as a whole. This idea was put into practice by several companies, e.g. Jean-Baptiste André Godin developed his factory for stoves and heaters, in Guise, northern France, in the same way during the 1840s.

In 1869, Jacob Cornelis van Marken (1845–1906) set up a factory in Delft, the Netherlands, for the production of yeast and methylated spirits. He encouraged employees to save and arranged insurance to cover them for illness, accidents and death. He made employees joint-owners of the company by

2 Sociaal-Economische Raad, *De Winst van Waarden* (Advice 00/11; The Hague: SER, 2000).

3 www.wbcsd.ch, 17 October 2005.

4 www.mdx.ac.uk/www/study/SHE8.htm#Owen, 17 October 2005.

giving them shares and installing a workers' council. Employees even received a few paid holidays. Van Marken also built houses for his workers to rent in a specially designed park where he also lived with his wife. However, this drew criticism as being too paternalistic.[5]

Similar communities were created in other countries, especially in the countryside. However, the motives of their creators were not always to contribute to society; often, it was mainly to keep control of the workers. In rural areas, workers were less influenced by socialism and trade unions, and could be paid lower wages. Social organisation of the factory-colony[6] guaranteed the workers' dependence on, and obedience to, the owner.

From pollution prevention to sustainable development

The US company 3M was a front-runner on environmental issues. In 1975, 3M set up its **Pollution Prevention Pays** (3P) programme to reduce costs by preventing waste and emissions—an innovative concept at that time. The programme helps to prevent pollution at the source (in products and manufacturing processes) rather than removing it after it has been created.[7] Projects have to meet the following criteria:

- Eliminate or reduce a pollutant

- Benefit the environment through reduced energy consumption or more efficient use of manufacturing materials and resources

- Save money through avoiding or deferring the need to invest in pollution control equipment, reducing operating and material costs, or increasing the sales of an existing or new product

A special award was introduced for projects that demonstrated technical innovation. Since 1975, 3M employees worldwide have initiated 4,973 projects under the 3P programme and, up to 2002, 3P had prevented emissions of 857,282 tons of pollutants and saved $894 million.

The 3P principle can be applied far more than is often assumed. If companies start searching for the sources of pollution, they often find unexpected possibilities for innovation. They also find that pollution prevention reduces the cost of:

---

5 International Institute of Social History, www.iisg.nl/collections/vanmarken/intro.html, 17 October 2005.

6 For example, Colonia Vidal in Catalonia (now a Museum of Technology History); www.museucoloniavidal.com.

7 www.3m.com/about3m/sustainability/policies_ehs_tradition_3pEx.jhtml, 17 October 2005.

- Maintenance

- Raw materials

- Accidents

- Clean-up operations

- Insurance

A Dutch project that took place between 1988 and 1991 aimed to exchange pollution prevention ideas between companies. This project showed that significant reductions in emissions could be achieved which were, at least, cost-neutral.[8]

Many companies have discovered the benefits of 'pollution prevention pays'. The originator of the idea of 3P was the chemical engineer Joseph Ling, who said that the environmental challenge for the next millennium 'is to move from pollution prevention toward sustainable development and design for the environment'.[9] This means that it is no longer sufficient to produce without pollution, but that the product should also be redesigned and even whole systems changed to fulfil the needs of customers in the most sustainable way.

The business principle behind sustainable development is often expressed as the **People–Planet–Profit** principle, or Triple P. A company has to find a good balance between these three aspects. Optimising for just one of the Ps leads to problems and is not acceptable.

- The **Planet** aspect means reaching a balance between the environmental burden and the capacity of the Earth to carry environmental burdens

- The **People** aspect deals with the communities and workers who have a stake in corporate activities—reducing poverty, good working conditions, contributing to community activities and community development, integration of immigrants, supporting democracy, etc.

- The **Profit** aspect implies that all economic activities must create prosperity for the company as well as its workers and owners. It is important to distinguish short-term and long-term perspectives (see above)

Embedding sustainability within a company requires investment in terms of time and money. Very often, it is not possible to determine exactly if and

---

8   S.C de Hoo, *Aanbevelingen voor een preventief milieubeleid* (The Hague: Rathenau Instituut, 1991).
9   P. Sorenson, 'Pollution Prevention Pioneer', *Inventing Tomorrow*, 1999, www.itdean. umn.edu/news/inventing/1999_Spring/ling.html, 17 October 2005.

when sustainability investments will pay back. Returns may occur in three ways:

- Economic value creation for the business in terms of product performance and production costs—this can be both short-term and long-term

- Value creation by improvements to the company's reputation and image—not only externally important but also internally as the motivation of personnel is influenced

- Value creation by increasing the coherence of various parts of the company and increasing their effectiveness and flexibility[10]

In addition, there is the risk of not embedding sustainability in the company.

As a result of better education and more wealth, people have become more emancipated.[11] Nowadays, individual preferences and opinions are more easily reflected in product choice, and verdicts about products and companies are made quickly.

Companies have to find a response to increased public pressure regarding sustainable development. The power of the consumer makes even the largest multinational vulnerable, as illustrated by what happened to Shell in the Brent Spar case.[12]

In 1995, the UK government allowed Shell to sink an old oil production platform in the North Atlantic. Greenpeace objected and tried to stop it by occupying the platform, but was removed. This event received widespread media coverage and Shell's petrol stations throughout Europe became deserted places. Various politicians joined actions against Shell. Ultimately, Shell changed its plans and the Brent Spar was dismantled and re-used in Norway.

There are various other examples of the effectiveness of consumer protest; the conditions under which chickens are kept were improved and seal hunting was curbed as a result of consumer protests.

10 J. Cramer, *Ondernemen met hoofd en hart, Duurzaam ondernemen: Praktijk ervaring* (Assen, the Netherlands: Van Gorcum, 2002).

11 For example, participation in higher education in Europe grew by a factor of more than three between 1970 and 1996. Unesco Institute for Statistics, www.uis. unesco.org/ev.php?URL_ID=5187&URL_DO=DO_TOPIC&URL_SECTION=201, 17 October 2005.

12 The Brent Spar affair is well depicted in a documentary ('Een geschenk uit de hemel: De slag om de Brent Spa') produced by Kees de Groot van Embden for VPRO-Television, Hilversum. For Shell's Brent Spar dossier, see www.shell.com/home/ Framework?siteId=uk-en&FC2=/uk-en/html/iwgen/leftnavs/zzz_lhn2_6_1.html& FC3=/uk-en/html/iwgen/about_shell/brentspardossier/dir_brent_spar.html, 17 October 2005. Greenpeace information is given at archive.greenpeace.org/ comms/brent/brent.html, 17 October 2005.

Now that companies are able to move (parts of) their business more easily from one country to another,[13] government regulations have become less important; in contrast, reputations and the public image of a corporation have become more important. Companies cannot deal with their responsibilities only by complying with the law. They have an increased responsibility, whether they like it or not.

Not only multinational companies, but also medium-sized and smaller companies have to deal with sustainable development, even in hard times. The bankruptcy of companies that cannot meet today's standards cannot always be prevented. This is illustrated by the following example.

HYTPSA

The major Spanish textile company Hilados y Tejidos Puigneró S.A. (HYTPSA) in northern Catalonia was founded in 1956. In 1996, its chairman and majority shareholder, Josep Puigneró de Sargatal, was the first Spanish businessman to be imprisoned for environmental violations. HYTPSA polluted the River Ter with toxic waste and had already been fined several times. Puigneró was also fined for employing 73 illegal immigrants. After 1996, a new spill meant that the legal battle continued. For almost ten years, HYTPSA management was engaged in legal battles. Meanwhile, business conditions were deteriorating as a result of cheap imports from Asia and, in 2000, HYTPSA went into temporary receivership.[14] Its workforce had already fallen from almost 3,000 a decade before to 700. In 2003, the European Commission decided that Spain had been giving the company illegal aid. This meant bankruptcy and only some minor company activities were maintained. The total debt was about €147 million.

National and international environmental laws have forced many laggards (non-proactive companies) to change their products or processes over time. This has resulted in less pollution. The HYTPSA example shows that spending large sums on lawyers and valuable management time on countering law enforcement is not a management strategy that shows great wisdom.

13 M. Castells, *The Power of Identity* (Oxford, UK: Blackwell, 1997).
14 *Troubled Company Reporter Europe* 1 (14 November 2000); bankrupt.com/TCREUR_Public/001114.mbx, 17 October 2005.

Embedding sustainable development in a company

The business planning, operations and control cycle

Sustainable entrepreneurship requires critical reflection of the company's core values, policy principles and operational procedures.

The challenge for sustainable entrepreneurship is integrating 3P into the main activities of a company. A sustainable substitute should be developed and implemented in each of the main functions of a company.

Implementation should be planned and evaluated carefully. It should be integrated in the ordinary planning cycles of the company. Figure 7.1 shows the different aspects of the business planning, operations and control cycle in a company.

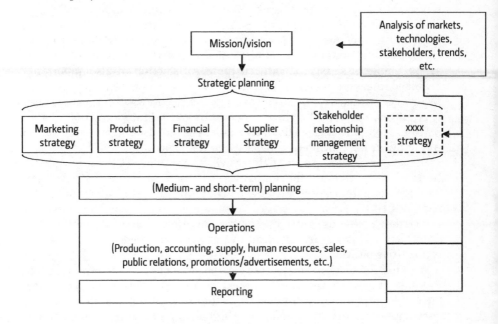

FIGURE 7.1 Business planning, operations and control cycle

Companies generally have vision and mission statements, though these usually remain unchanged for at least a decade. New strategies are formulated based on these statements and on analyses of external aspects such as stakeholders and trends, and internal aspects such as financial and sales results. The general strategy of a company is often divided into strategies for specific areas, which are interdependent and may include:

- Personnel strategy (what knowledge do we want inside the company?)

- Product strategy (what type of products do we want to sell?)

- Marketing strategy (what customers do we want to appeal to?)

- Production strategy (quality, costs, flexibility, resource supply)

- Financial strategy (short term versus long term, capitalisation)

- Stakeholder management strategy (government, non-governmental organisations, local communities, transparency, dialogue)

Once the strategic planning is complete, operations are planned for the medium and short term. How many products will be produced? How many raw materials will be purchased? How many people will be hired? What skills are required? What type of research will be planned?

Planning is followed by operations and then by reporting. Reporting includes not only finance. Sales results, quality control results, employment, safety, employee satisfaction and customer satisfaction are also important.

Embedding sustainability into a company is a long-term process that requires continuous attention by management. But, as all companies need to be entrepreneurial, the process is never finished. Embedding sustainability in the company creates not only challenges but also threats—innovation might fail or have unforeseen negative consequences.

However, embedding sustainability will create an innovative atmosphere. Innovation is important for companies striving to maintain their competitive position. Innovation also means changing things in a company and changing things, if well managed, keeps employees alert and enthusiastic. Innovation also tends to attract other organisations that want to collaborate on projects. This in turn can lead to increased knowledge and a stronger competitive position.

Working on sustainability often has positive effects. For example, Gist Brocades (the yeast company founded by van Marken) developed a new anaerobic purification technology[15] in the 1970s in order to deal with an enormous problem with its effluent. This technology was subsequently marketed and became the core of a new company: Biothane.

The tools or principles of sustainable development are important for each aspect of the business planning cycle. A selection of these is discussed below.

15 www.biothane.com/default.aspx?sel=31, 11 March 2006.

Sustainable mission statements

Many large companies such as DaimlerChrysler,[16] Swiss Re,[17] Shell,[18] Unilever[19] and Ford,[20] but also smaller companies such as confectionery manufacturer Perfetti-van Melle,[21] have incorporated sustainable development into their mission statements.

Over 120 companies signed the declaration for sustainable development during the World Industry Conference on Environmental Management in 1991. Others have signed the CERES (Coalition for Environmentally Responsible Economies) principles. CERES[22] is a community of forward-looking companies (e.g. Nike, American Airlines), which have committed themselves to continuous environmental improvement by endorsing a ten-point code of environmental conduct.

Incorporating sustainable development in mission statements ensures that top managers remain committed. They are the ones who formulate the mission and stay committed to it by communicating it time after time to their employees and external contacts.

A company that is just starting up can be flexible when creating a sustainable mission statement, as it has not yet invested in a specific product, process, technology or market. It can still decide who will be its suppliers, retailers and consumers. It can choose suppliers and retailers that maintain a high standard of sustainability in their part of the supply chain. The company might also focus on customers that value sustainability.

A proactive sustainable company has a major advantage over 'dirty' laggards. It can work from its own vision and hardly needs to react to government measures or the conditions of environmental licences as the company is often seen as the benchmark for these legal measures. For example, European farmers who have made the transition to organic farming will not be confronted with further tightening of regulations governing the use of pesticides and the treatment of animals.

Stakeholder relationship management

Stakeholder relationship management concerns all the company's internal stakeholders (e.g. specific groups of employees and management) and its

16 www.daimlerchrysler.com (go to 'Sustainability [Economy, Environment, Social]'), 17 October 2005.
17 www.swissre.com (go to 'About Us', then 'Corp. responsibility' and then 'Sustainability'), 17 October 2005.
18 www.shell.com (go to 'About Shell' and then 'Our Strategy'), 17 October 2005.
19 www.unilever.com/ourcompany/aboutunilever/default.asp, 17 October 2005.
20 www.ford.com/en/company/about/corporateCitizenship/report/default.htm? referrer=home, 17 October 2005.
21 www.perfettivanmelle.nl/english_site/melle/index.htm, 17 October 2005.
22 www.ceres.org, 17 October 2005.

external stakeholders such as customers, suppliers, local communities and non-governmental organisations (NGOs).

Careful stakeholder relationship management can make the difference for the success of sustainable entrepreneurship. Many companies still regard stakeholders as hostile forces that need to be appeased. However, treating stakeholders as partners and sources of knowledge is often a more successful attitude to adopt.

Stakeholder management exists in addition to traditional public relations management. One of the reasons why it is so important is the increased vulnerability of companies to changes in public opinion. A proactive attitude to monitoring trends in society means that sudden negative publicity can be avoided. A good 'social antenna' is crucial to foreseeing what society will consider right or wrong and how new products will be received. Interaction with stakeholders is useful for product development, marketing, the formation of strategic alliances and the development of general company policies. Best practices include:

- Create an open organisation with internal and external dialogues
  - Exchange and discuss points of view, problems and dilemmas
  - Consult consumers and NGOs when new strategy is formulated. NGOs have an antenna for the opinions and feelings of consumers
  - Communicate in time to reduce the risk of mistakes; e.g. the battle mentioned above between Shell and Greenpeace on Brent Spar was partly based on mistrust and lack of dialogue

- Seek support from external stakeholders. For example, WWF supports and quality-controls the 'green electricity' supplied by NUON. This increases the reliability and credibility of its activities

- Report transparently about the company's activities and the consequences of these activities. Good reports also determine a company's image

- Provide consumers with relevant background information about products

The many different stakeholders will need to be approached separately. Sustainable energy, for example, can be positioned as 'good for the environment' or 'good for the future of your kids', but also as a 'technical innovation' or a 'new business for investment'.

Social, environmental and sustainability annual reports

Companies publish annual financial reports and usually quarterly results (see below). Many companies publish an annual social report which includes information regarding its employees, pensions, occupational safety and health records, etc.

Over the last few years, a growing number of companies have started to publish annual environmental reports and sustainability reports. These reports include information about the company's environmental performance and sustainability activities (e.g. its behaviour in developing countries).

An initiative that is worth mentioning here is the Global Reporting Initiative (GRI). [23] GRI is an independent institution whose mission is to develop and disseminate globally applicable sustainability reporting guidelines. These guidelines are for voluntary use by organisations to report on the economic, environmental and social dimensions of their activities, products and services. The GRI incorporates the active participation of representatives from business, accountancy, investment, environmental, human rights, research and labour organisations from around the world. Started in 1997 by CERES, the GRI became independent in 2002 and is an official collaborating centre of the United Nations Environment Programme (UNEP).

The GRI is developing a globally applicable framework for reporting an organisation's sustainability performance. The framework presents reporting principles and specific content indicators to guide the preparation of sustainability reports. Within this framework, companies report in the same format on indicators such as water consumption and energy consumption, but also on human rights programmes undertaken. Applying this framework allows the environmental performance of different companies to be compared.

Sustainability and marketing

Sustainability marketing concepts

Sustainability in marketing began with 'environmental product marketing' (i.e. introducing a product with an environmentally sound component to the market). This was followed by cause-related marketing (i.e. adding a new layer of values to a product or brand by connecting it with a social goal or societal organisation).

New marketing concepts that are linked to the People–Planet–Profit principle focus on the needs of consumer and (the well-being of) society. A definition for sustainable entrepreneurship from the marketing perspective is as follows:

> The concept of sustainable entrepreneurship holds that the organisation's task is to determine the needs, wants and interests of target markets and to deliver the desired satisfactions

23   www.globalreporting.org, 17 October 2005.

more effectively and efficiently than competitors in a way that preserves and enhances the consumer's and society's well-being.[24]

Besides choosing the right product or, in other words, fulfilling the right need, marketing includes:

- Pricing (what should be the cost of a product that contributes to sustainability?)

- Distribution (what type of shop will sell my product?)

- Promotions (what product characteristics will be communicated and how?)

Motives for including sustainability in marketing have to do with the image/reputation of the company and its process of business value creation. More specifically, these motives are:

- Companies that characterise themselves as 'caring' have more loyal customers. People choose products that contribute to sustainability for emotional or ethical reasons. A choice based on these grounds is more permanent than a choice based on price/performance

- Studies indicate that reputation is responsible for a considerable part of the total value of a company. A few numbers to illustrate this:[25]
  - A 60% improvement in reputation explains a 7% increase in stock value
  - About 40% of the market value of a company is linked to its reputation
  - A reputation crisis leads statistically to a stock value decrease of 8%

- Consumers want to have a good feeling about the things they buy. But they also require convenience and they want to be addressed personally. This is one of the reasons why organic food shows good sales. It gives consumers the feeling that the producer cares personally about their health (no luxury in times of scandals like chemical waste in animal food, mad cow disease and salmonella in chicken)

24 P. Kotler, *Marketing Management: Analysis, Planning, Implementation and Control* (Englewood Cliffs, NJ: Prentice Hall, 7th edn, 1991).
25 J. Cramer, *Duurzaam in Zaken* [*Sustainable in Business*] (inauguration speech, Erasmus University, Rotterdam, 2 February 2001; Assen, the Netherlands: Van Gorcum, 2001).

- Products that contribute to a sustainable future have an enormous growth potential. It is a challenge to develop products that trigger customers' consciences. If successful, and the new product is more than only a technical or cosmetic improvement of an existing product, it will be harder for competitors to copy it

- Consumers often mistrust advertising. Independent information from government agencies or NGOs is better for winning the trust of customers than advertisements

All this may suggest that sustainable marketing is very attractive and has only advantages. However, the following also need to be taken into account to be successful in sustainable marketing:

- Ensure that the sustainable development approach has a link with the company's roots. This will boost internal support and avoid customers turning their back on products because the company's image has changed too drastically

- The internet generation has a healthy distrust towards advertising and a strong intuition for commercial influences. This has led to a critical attitude towards companies and brands. Many believe that companies and brands have too much influence on people's personalities

- Consumers are not consistent. All want to save the environment, but they still want to holiday on a far-away beach, preferably two or three times a year

- Communications have to first target internal customers (especially employees), so that they appreciate what sustainability means to their company. Employees are, after all, the ambassadors of the enterprise and need to have a positive attitude towards sustainability. It is vital that the highest level of management supports sustainability as it starts by embedding it at the strategic level of business planning. This sets an example for the rest of the company

Eco-labels

Environmentally conscious consumers often find it difficult to buy the products they want. Because every company could advertise its products as 'environmentally sound', eco-labels have been developed by government agencies, industrial sectors and environmental NGOs to help consumers select the 'right' product. Figure 7.2 shows some examples.

The Energy Star® programme (Figure 7.2a) promotes energy efficiency and is probably one of the best-known and most successful eco-labels. It was established in 1992 by the US Environmental Protection Agency and was intended at first only for computers (mainly to introduce sleep mode). Its enormous success means that the programme now labels more than 35 prod-

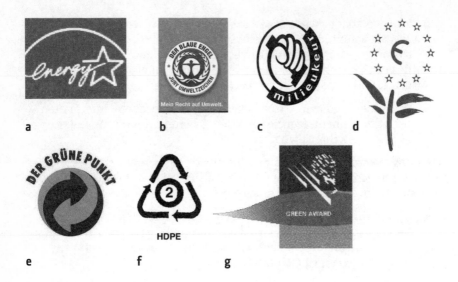

FIGURE 7.2 Examples of eco-labels

uct categories including houses. More than 1 billion Energy Star products have been sold in the USA alone.[26]

The organisations responsible for the official German 'Der Blaue Engel' (Blue Angel)[27] (Figure 7.2b) and Dutch 'Milieukeur' eco-labels[28] (Figure 7.2c) produce lists of environmental and quality requirements for each product group. These requirements are based on the complete product life-cycle and are stricter than those for the official EU eco-label (Figure 7.2d).[29] This was introduced in 2000 and has been developed for a large number of product groups. It is intended to replace many other labels. Lists of criteria for each product group have to be met. These criteria are based on a consensus established between Member States.

A product bearing the German 'Der Grüne Punkt' ('Green Dot'; Figure 7.2e) means that the producer has paid for the processing of the packaging and that this is collected separately. The logo is valid only in Belgium, France, Germany and Austria. The logo is not an eco-label, as it does not guarantee the packaging will actually be recycled.

26 www.energystar.gov/ia/partners/downloads/energy_star_report_aug_2003.pdf, 17 October 2005. S.O. Andersen and D. Zaelke, *Industry Genius* (Sheffield, UK: Greenleaf Publishing, 2003).

27 www.blauer-engel.de/englisch/navigation/body_blauer_engel.htm, 17 October 2005.

28 www.milieukeur.nl, 17 October 2005.

29 europa.eu.int/comm/environment/eco-label, 17 October 2005.

Symbols like the one in Figure 7.2f for high-density polyethylene (HDPE) are often placed on plastic products made from a particular polymer. The number or the letters inside indicate the type of plastic. The symbol means that the product could technically be recycled. Actual recycling requires separate collection of the same type of plastic.

Some sectors use specific labels. For example, the Green Award (Figure 7.2g) is available for crude oil tankers, product tankers and bulk carriers to apply for. In return for compliance with environmental and safety standards, Green Award vessels receive an incentive based on port dues in a number of countries.[30]

Numerous companies have chosen to use these labels and to produce products according to their rules because they realise that they add an extra selling point to their products.

There are many eco-labels worldwide. The Global Ecolabelling Network[31] fosters co-operation, information exchange and harmonisation among eco-labelling programmes.

In some cases, product symbols are made up entirely by company's marketing departments. They attempt to give customers the feeling that the product is better for the environment, without necessarily being true. Unfortunately, there is no mechanism to control this practice.

Production

Environmental management systems

Historically, companies have been confronted with various regulations regarding their emissions. These problems have always been closely related to living and working conditions at and around production facilities.

Before the 19th century, there was a long history of local regulations. Many cities prohibited dumping of waste in streets, canals or rivers. In the Netherlands, the first national law in this respect was the Factory Law of 1875, which contained provisions regarding nuisances such as noise and odour. This law typically created 'triple D' solutions (see also Chapter 10):

- Dumping (in pits)
- Displacement (waste outside city, chimneys, etc.)
- Dilution (in rivers, etc.)

However, simple answers were no longer satisfactory once it was no longer just waste that was criticised but also excessive resource consumption. A

30 www.greenaward.org, 17 October 2005.
31 www.gen.gr.jp/index.html, 17 October 2005.

company's organisation was affected. Environmental management gradually emerged into a sophisticated environmental management system (EMS).

Environmental management systems of today match perfectly with the trend to implement quality control systems such as those published by the International Organisation for Standardisation (ISO)[32] and the Hazard Analysis Critical Control Point (HACCP)[33] method in food production in the 1980s and 1990s. In brief, ISO and HACCP require:

- The writing-down of business policies, processes and organisations
- Work carried out according to what has been written
- Ability to avoid and trace mistakes

An EMS is a method designed to systematically improve the environmental performance of an organisation. It addresses the immediate and long-term impacts of products, services and processes on the environment and is embedded in the organisation's overall management structure. Results are achieved through the allocation of resources, assignment of responsibility and ongoing evaluation of practices, procedures and processes.

In the 1980s, more and more organisations systematically paid attention to environmental matters with the aim of reducing pollution and minimising energy and raw material consumption. Guidance was published to help companies setting up their own EMS. Environmental management systems were further developed in the early 1990s when ISO published its ISO 14001 standard leading to certification for an EMS.

An EMS consists of several elements:

- Environmental policy formulation, which includes specific goals
- Environmental programme, which includes the intended activities
- Integration of environmental care in operational management
- Internal information service, education and training
- Internal control, measurements and registrations
- Internal and external reporting, e.g. through an environmental annual report

The methods discussed here are not limited to industrial enterprises. Environmental management systems are important for all kinds of organisations such as retailers, the catering industry, insurance companies, banks, governments, hospitals, schools, consulting engineers, etc.

However, successful environmental care requires some essential conditions to be met:

32 www.iso.ch, 17 October 2005.
33 International HACCP Alliance, haccpalliance.org, 17 October 2005.

- A clear vision by top managers with respect to the positioning of environmental care in the overall business policy

- Awareness of the importance of environmental care at all layers and in all parts of the company

- Availability of time, people and means for environmental care

- Integration of environmental care in all business processes with a direct or indirect environmental impact

Measures should be aligned carefully with other areas of care such as quality control, occupational safety and health, and reward and social care systems.

Environmental management can be supported by various technologies. Examples include:

- Time switches that shut off water taps, switch off lights or put monitors to 'sleep'

- Fuel-efficient vehicles and transport planning

- Detection of hydrocarbons using organic vapour analysers (important in chemical and petroleum processing)

- Use of sensitive acoustic equipment to detect leaks from underground pipelines (they measure the noise produced by leaks)

The key element in various measurements is to motivate people to contribute to improved environmental performance. Feedback (e.g. regarding less waste or reduced energy use) is therefore very important. Software that registers performance also helps to monitor and motivate people. It can also warn the organisation of deteriorating performance.

End-of-pipe solutions

Environmental management systems are designed to avoid emissions and save resources; they are not enough so solve all environmental problems.

Some waste is inherent to a production process. End-of-pipe solutions are necessary to prevent this waste being discharged or emitted untreated and thus harming the environment. In industry, end-of-pipe solutions are clean-up devices installed after the core production units.

End-of-pipe technologies offer a number of advantages:

- They do not necessitate change to the productive process. Any change to the core production process involves risks for product quality and continuity of production. It often requires considerable investment and retraining of personnel. Companies therefore tend to avoid changes to their production processes.

- Capital and operating costs are foreseeable and controllable. It may be possible to order the equipment 'off the shelf' and to put its operation out to contract.

However, waste-streams can pose demands on end-of-pipe technologies, e.g. the volume and concentration of hazardous materials in waste-water entering a biological treatment plant. It is therefore often financially attractive to minimise waste-streams before they are treated and/or build storage facilities to ensure their efficient operation. Waste-water treatment is a typical example. The volume of waste-water can often be reduced by simple means, but this may involve changes to production processes.

End-of-pipe technologies can be divided into four categories:

- **Separation** technologies that separate harmful from harmless substances, e.g. filtration, precipitations, condensations, etc.

- **Treatment** technologies that eliminate harmful substances, e.g. flaring off gases, oxidation, etc.

- **Shielding** (from noise, radiation, light)

- **Compensation**: new nature reserves to compensate for the damage from building new industrial sites, forests to trap carbon dioxide (CO_2) emissions, wildlife tunnels under motorways, etc. These leave the source unaffected

Beginning in the early 1970s, various countries introduced new legislation that resulted in rapid growth of these end-of-pipe solutions. Purifying water before discharging it, fitting dust filters in chimneys and building noise protection walls are some of the actions taken. Although they do not generally solve the problem completely, end-of-pipe solutions are still common when no alternative is available.

Supply chain management and industrial ecology

Any organisation tends to optimise its performance within its own limits. However, this is often not the optimum for society. For example, a company that is emitting too much CO_2, nitrogen oxide (NO_x) and soot by burning fossil fuels to heat its processes might decide to change its processes and use electricity instead. This will solve its emissions problem, but probably offers no improvement for the environment. A power plant will have to supply the electricity, but its efficiency (power plant, transmission network, etc.) will probably be 25–50% (i.e. lower than most boilers); the environment is not protected by this change. A large power plant may clean its flue gases better than a small or medium-sized enterprise (SME), but the change is unlikely to be beneficial for the environment.

Especially if we target emissions that are harmful from a global or continental perspective, we should therefore consider the whole supply chain.

The supply chain covers all the steps from exploration of basic raw materials and resources to the end-of-life disposal (or recycling) of a product.

Decisions at specific points in the supply chain may have important consequences at other points. For example, material recycling might be seriously hampered by the presence of small amounts of additives (e.g. plastic coatings ruin the recycling of cardboard). It is thus important to take product and process decisions within the perspective of the whole supply chain.

Industrial ecology

In 1989, Frosch and Gallapoulos[34] suggested that industries should be more efficient with their material and energy flows and should resemble ecological systems. Ecology served as an example for the new order in which industrial systems should be modelled.

Industrial ecology is a useful approach for the supply chain problems outlined above. The meaning of 'industrial ecology' has varied over the years, but the most important definitions are those of Graedel and Allenby and the Institute of Electrical and Electronic Engineers (IEEE).[35] In 1995, Graedel and Allenby proposed:[36]

> Industrial ecology is the means by which humanity can deliberately and rationally approach and maintain a desirable carrying capacity, given continued economic, cultural and technological evolution. The concept requires that an industrial system be viewed not in isolation from its surrounding systems, but in concert with them. It is a systems view in which one seeks to optimize the total materials cycle from virgin material, to finished material, to component, to product, to obsolete product and to ultimate disposal. Factors to be optimized include resources, energy and capital.

In the same year, the IEEE added:[37]

> Industrial ecology is the objective, multidisciplinary study of industrial and economic systems and their linkages with fundamental natural systems. [. . .] It is important to emphasize

34   R.A. Frosch and N.E. Gallopoulos, 'Strategies for Manufacturing', *Scientific American* 261.3 (1989): 94-102.
35   Institute of Electrical and Electronics Engineers (IEEE), Environment, Health and Safety Committee, *White Paper on Sustainable Development and Industrial Ecology* (New York: IEEE, 1995).
36   T.E. Graedel and B.R. Allenby, *Industrial Ecology* (Englewood Cliffs, NJ: Prentice Hall, 1995).
37   Institute of Electrical and Electronics Engineers (IEEE), Environment, Health and Safety Committee, *White Paper on Sustainable Development and Industrial Ecology* (New York: IEEE, 1995).

that industrial ecology is an objective field of study based on existing scientific and technological disciplines, not a form of industrial policy or planning system.

In the last few years, the concept of industrial ecology has evolved from just an interesting term for environmental measures in industry to a field of study aiming at environmental, organisational, economic and technological renewal of industrial society. Industrial ecology is a growing discipline.[38]

The difference between industrial ecology and designing a single product or process for sustainability (for instance, using life-cycle analysis; see Chapter 8) is that a complete industrial sector or supply chain is considered rather than looking at just one product. All material flows (resources, energy and capital) within that supply chain are considered. Industrial ecology studies can become quite extensive. Figure 7.3 provides a schematic overview of the material flows to be mapped.

Industrial ecology maps flows of materials going in and waste flows going out. Several supply chains present on the same industrial site can exchange material or energy for further optimisation. A famous example of industrial ecology in practice is Kalundborg in Denmark (Figure 7.4). This complex of various industrial facilities minimises inputs of energy and virgin materials by exchanging energy and waste/raw materials.

The people 'pillar' of sustainable development in companies

The social (people) dimension of sustainable entrepreneurship is about the consequences for and conditions of people inside and outside the company. It is about:

- Working conditions and education within the company
- People living in the neighbourhood
- Users of the products and/or services
- People living in developing countries where production units are located

In summary, it is about people in the whole supply chain.

For people inside the company, it is expressed as 'employee satisfaction'. In the first place, this depends on:

- Physical working conditions
- Social and cultural working conditions

38   International Society for Industrial Ecology, www.is4ie.org, 17 October 2005.

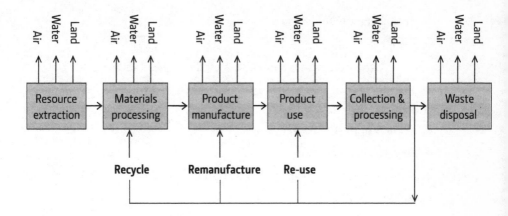

FIGURE 7.3 **Material flows**

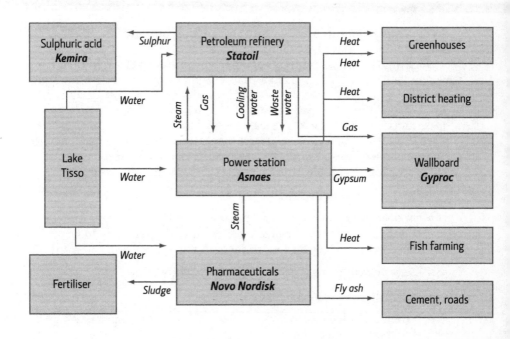

FIGURE 7.4 **Schematic of Kalundborg industrial site**

Source: 'Industrial Symbiosis at Kalundborg, Denmark', www.indigodev.com/Kal.html, 17 October 2005

- Terms of employment
- Fringe benefits

However, employee satisfaction also includes aspects such as:

- Membership of a union and the availability of a trusted representative
- Privacy of information
- Registration of email
- Possibilities for education and training
- Regular feedback
- Equal treatment
- Availability of a social plan in case of reorganisations and conflicts
- Company healthcare
- Provision of a day nursery

The company culture and being involved in the policy-making of the work unit or company may also be important factors.

Working conditions and terms of employment have a direct influence on worker satisfaction. They include the location of the work (e.g. a large, small, dark or light room with or without air-conditioning), salaries and profit share. Employee satisfaction should be checked regularly, e.g. via questionnaires. It can also be measured by looking at the number of employee initiatives such as proposals for innovation within a company, excursions or office parties. A high rate of absence is a warning sign.

Consideration of people external to the company includes issues such as:

- Respect for human rights
- A clear policy about doing business in countries where human rights are seriously violated and, eventually, standards to be applied for its own operations in these countries
- Selecting suppliers according to sustainability criteria
- A ban on the use of child labour
- Stimulating local economies and social activities (in places where the company operates)
- Stimulating local education (which might result in a better-trained workforce)

Supporting local democracy might diminish conflict in local communities. Sponsoring of beneficial organisations may improve social life. Direct measures like reforestation might sustain the livelihood. The key for success, however, is to refrain from any form of paternalism. Local communities set their own priorities. For example, Shell makes the following statement about its operations in the Niger Delta:[39]

> Today our community development program in the Niger Delta region is based on the principles of sustainable development and best global practice. In 2001 we invested over $50 million in health, education, agriculture, job creation, women's programs, youth training and sponsorship. Everything we do is guided by expert advice from our stakeholders and strategic partners and increasingly open and honest communication with the communities.

The social impact of using the products as sold by a company is another side of the people pillar of sustainable development. What is the social impact of a product? Is research and production acceptable? To assess the social impact of technology, see Chapter 6.

Accounting methods for sustainability

Creating proper financial reports about sustainability activities within a company is not easy. One can report about environmental aspects and financial results but, in most cases, the negative consequences of unsustainable activities are not accounted for. In addition, indirect profits of sustainability activities (e.g. extra productivity) are difficult to measure and little is known about social indicators.[40] In the absence of international standards, benchmarking between companies is rarely possible.

Although many companies have shown that a sustainable way of working can be combined with being profitable, many others still think it will cost money and weaken their competitiveness. This belief is caused by the ignorance of many entrepreneurs about the potential savings that sustainable products, processes and technologies can lead to.

Environmental economists give the following reasons for not introducing profitable environmental measures:[41]

39 www.shell.com, 17 October 2005.
40 J. Ranganathan, 'Signs of Sustainability: Measuring Corporate Environmental and Social Performance', in M. Bennett and P. James (eds.), *Sustainable Measures: Evaluation and Reporting of Environmental and Social Performance* (Sheffield, UK: Greenleaf Publishing, 1999): 475-95.
41 S. Schaltegger and K. Muller, 'Calculating the True Profitability of Pollution Prevention', in M. Bennett and P. James (eds.), *The Green Bottom Line. Environmental Accounting for Management: Current Practices and Future Trends* (Sheffield, UK: Greenleaf Publishing, 1998): ch. 3.

- Environmentally related costs are considered as overheads and therefore accounted to all products of the company instead of just being accounted to the polluting products

- Environmentally related costs are overlooked, as are the benefits of sustainable products

- Investments in sustainable products and processes are undervalued because financial evaluation techniques have time horizons that are too short

Environmental costs could be **internal costs** such as the cost of on-site water treatment. But they could also be **external costs** such as the cost to homeowners of cleaning the façade of their houses to remove the affects of passing traffic or the costs of healthcare if the health of local residents has been affected.

For sustainable development, it would be best if investment decisions were based both on internal and external costs. However, polluting companies often do not pay external costs. They frequently take account of the possibility that an external cost will become an internal cost, e.g. liability for damages such as the health effects of soil contamination.

There are three ways in which companies can improve the visibility of internal costs. These are:

- Environmental costs should not be considered as overhead costs

- Total cost assessment should be used to obtain a complete overview

- Investments should be based on an adequate time horizon

These three methods are discussed below.

Do not consider environmental costs as overhead costs

Overhead costs are defined as all costs that are not assigned to a specific process or product. To calculate the production cost of a product, companies usually distribute overhead costs over all products and processes. By separating environmental costs from other overhead costs and assigning these costs only to those products and processes that are actually causing them, the correct production costs of products can be determined.

Even if all products pollute about as much, the principle of regarding these costs as overheads discourages the managers responsible from taking steps to reduce the impact on the environment. The problem is that their budget has to bear the cost of reducing pollution but the whole company benefits from lower pollution costs.

Use total cost assessment to get a complete overview

The total cost assessment[42] approach was developed as a reaction to traditional accounting methods. It enables a more accurate calculation of product cost price and can be used to make environmental costs more visible and explicit. Coopers & Lybrand Consultants[43] proposed five debit areas:

- **Conventional:** costs for resources, utilities, capital and labour that are also accounted for by traditional methods

- **Hidden:** costs that are mostly invisible using traditional methods as they are regarded as overheads (includes costs that will occur in the future, though it is may not be clear when)

- **Contingency:** costs that the organisation is not sure whether it will have to pay in the future

- **Public relations (PR) and image:** costs that have to be made to achieve a 'good' image, but also the financial consequences of a 'good' or 'bad' image

- **Social:** external costs of pollution such as diminishing value of local property

As the numbers of these costs increase, so they become harder to determine. For each of these categories, an inventory should be made of costs and benefits of the current product and compared with the costs and benefits of a sustainable alternative. Some future costs and benefits may be hard to determine and a best estimate will suffice. Table 7.1 shows examples of the different categories.

After an inventory has been made of all these costs, they should be assigned to the product that is accountable for them. If more than one product is accountable, a fair distribution should be applied.

Evaluate investments based on an adequate time horizon

Any investment decision should be based on an assessment of all expected costs and benefits. The same applies to investments in sustainability. But conventional investment assessment methods undervalue sustainable alternatives.

---

42 A.L. White, M. Becker and J. Goldstein, *Total Cost Assessment: Accelerating Industrial Pollution Prevention through Innovative Project Financial Analysis* (Washington, DC: US Environmental Protection Agency, 1992).

43 Coopers & Lybrand Consultants, *On Environmental Management* 3.1 (1998); www.pwcglobal.com/extweb/service.nsf/8b9d788097dff3c9852565e00073c0ba/1211fe28 6669bba18525692e0004fad5/$FILE/EnvironmentalAccountingFeb98.pdf, 17 October 2005.

Cost/benefit category	Examples
Conventional	(+) Lower costs of resources (materials) and waste management
	(+) Lower costs because of reduced energy consumption in production processes and transport
	(+) Lower costs of capital because of subsidies for investment in environmentally friendly solutions
	(+) Higher selling price as the consumer is prepared to pay more for a sustainable product
	(+) Bigger market share
Hidden	(−) Sustainable product could trigger more investments in research and development
	(+) Savings in waste management costs and pollution monitoring costs
	(+) Savings in labour costs due to safer working conditions, lower sick leave rates and more motivated employees
	(+) Choosing sustainable energy could reduce the cost of dismantling conventional or nuclear power plants in the future
Contingency	(−) Fines because of a lack of permits, ultimately resulting in forced closure of the company
	(−) Costs of claims because of nuisance affecting others
	(−) Costs of adaptation of products or processes to comply with new environmental legislation
	(−) Costs of (future) charges and higher insurance premiums
	(−) Costs of cleaning up pollution following accidents
	(+) Benefits because the sustainable product is of higher quality and thus results in higher sales and lower after-sales costs and claims
PR and image	(+) Benefits of a 'green' image increases sales
	(−) Costs to propagate a 'green' message such as annual environmental reports
Social	(+) Better relations with neighbours
	(−) Message of 'improvement' can only be conveyed successfully if the previous worse situation is known

TABLE 7.1 Examples of the five cost/benefit categories

Payback period or payout time is often used to evaluate investments because it is easy to use. This investment assessment method considers the total costs of capital expenditure at start-up and the expected annual benefits. Dividing the capital cost by the expected annual benefits results in the payback period. The time horizon of this method is usually short and the longer-term benefits of sustainable investments are often not taken into account. One example is the huge liability claims common in the USA for soil and groundwater contamination.[44]

Net present value (NPV) offers the possibility of taking future costs and benefits into account. This investment assessment method is a way of comparing the value of money now with the value of money in the future. A euro today is worth more than a euro in the future because inflation will erode its purchasing power, while money available today can be invested and grow. For example, assuming a discount rate of 5%, the NPV of €2,000 ten years from now is €1,227.83. So if someone offered you €1,000 now or €2,000 ten years from now, you would pick the latter because its NPV is higher. If the NPV of an investment is positive, the project is regarded as profitable.

Sometimes, projects are of such strategic importance for a company that they are carried out despite a negative NPV. These projects can offer the company new opportunities or prevent it from being forced out of business.

Investment assessment methods often calculate the payback period for the life expectancy of equipment assuming that its value has fallen to zero. However, longer use or a considerable value left after the normal life expectancy can create considerable profits.

Questions, discussion and exercises

1. Evaluate the business strategy of Interface (see over) and compare it to one of its competitors in the floor coverings market. What are significant differences? How could Interface proceed towards sustainability?

2. Same question for CONSTRUCCIONES RUBAU S.A. (see page 197), formed as an 'anonymous society' (board of directors) in 1968 in Girona, Spain.

3. Estimate the NPV of €1 million invested in a 1 MW wind turbine 15 years from now. Compare it to buying bonds with a 6% interest rate.

   a. The investment is €1 million, the annual average electricity production is 2.1 GWh, the annual maintenance costs are €20,000 and the electricity is sold at €0.04/kWh.

   b. What will be the NPV if energy becomes scarce and the price of this green electricity rises to €0.08/kWh?

4. Compare the sustainability reports of two competing large companies in the same sector. How do these companies differ? Which company contributes most to sustainable development in your view?

---

44 US Environmental Protection Agency, 'Superfund: Cleaning Up the Nation's Hazardous Waste Sites', www.epa.gov/superfund, 17 October 2005.

For example, compare: Shell and BP or ExxonMobil; Du Pont and BASF; DaimlerChrysler and Toyota; Boeing and Airbus; Procter & Gamble and Unilever; Nike and Levi's; Sony and Philips.

CONSTRUCCIONES RUBAU S.A. have had their Environmental Management System certified since the year 2000 and covers all types of works. This system also integrated in the CPM system is one of the fundamental axes of the environmental policy of **CONSTRUCCIONES RUBAU S.A.**

Further afield even than policy or legislative requirements, **CONSTRUCCIONES RUBAU S.A.** heads a campaign called **'GREEN CONCRETE'** which claims the optimisation of environmental behaviour in the works through a more efficient implementation of all those involved in the works. Among the measures which have been developed, important to note are:

- **Works posters**. Editions of informative posters which through caricature indicate the best environmental practices.

- **Green Guide**. A pocket guide to deliver to all workers of our company and management of subcontractors, which contains the necessary information to administer the work correctly from an environmental point of view.

- **Waste bin especially for dangerous residues**. Waste bin designed for the various dangerous residues generated by a work, thanks to its compartmentalisation.

- **Prize for the best environmentally friendly subcontractor**. An award which is given every year to the subcontractors of **CONSTRUCCIONES RUBAU S.A.** to show that they have shown best practice from the point of view of the environment.

- **Involvement of society**. The involvement of **CONSTRUCCIONES RUBAU S.A.** with society through speeches, work sessions for teaching visits, participation in associations for the defence and protection of the environment etc.

- **Promoting environmental awareness of subcontractors**. Imparting environmental awareness to the subcontracted companies is basic in order to claim that the works achieve their environmental goals.

- **Encouraging technological innovations for the environment**. It is important that the company assumes a firm commitment to look for new formulae which will allow for building in a more sustainable form. A clear example is the recycling of the asphaltic hot bituminous mixes.

Source: www.rubau.com/eng/ap9_i.htm

8 Design and sustainable development

Engineers have various roles in their professional life as:

- Project managers
- Plant managers
- Technological researchers
- Inspectors of technological systems

But one of their most important tasks is the methodological design of products, processes and systems. Engineers are able to optimise the design using their knowledge of scientific principles.

Decisions are made in the design process that have a considerable effect on the sustainability of the product, process or system. Engineers can also be involved in the design process as a representative of the organisation that commissions the design or the organisation that will use it.

In this chapter, we deal first with designing as part of an innovation process. Later we focus on designing in detail.

What is designing?

Design within the innovation process

Traditionally, the design process is seen as one link in a chain of steps. An organisation commissions a new process or product to help it reach its goals. The engineer that accepts the task of delivering the design is given a programme of requirements, which the product should ultimately adhere to. This list of requirements is also used subsequently to evaluate the design that the engineer has produced. Once the design is accepted, the job of the designing engineer comes to an end.

This traditional view of engineering design is characterised by three activities:

- Commissioning
- Design
- Application

The programme of requirements or specification separates the design process from the commissioning and the application process. It therefore has paramount importance—the designer is neither supposed to question the nature of the design nor its application.

Once the designer has accepted the programme of requirements, the next step is to obtain a solution that fits into all requirements completely (even if the demands seem to be inconsistent). This lack of critique depends on the engineer's culture. Western engineers are often somewhat reluctant to criticise the commissioner but, in more hierarchical cultures such as those in East Asia, a lack of courage to question the wisdom of the commissioner can drive engineers to months of useless work.

Interaction between designers and commissioners is very important, as it often leads to altered requirements. Sometimes the designer knows of possible ways to reduce costs or additional features of which the commissioner is unaware. The designer may also be able to offer valuable advice on the use of the product such as:

- How best to operate the product
- How to maintain it
- How to improve the design based on practical experience

Interaction between the designing engineer, the commissioner and the user benefits everyone involved with the product. A strict separation of designing from commissioning and application is a barrier to productive interaction and to responsible behaviour. It can also lead to ethical problems:

> Consider the Dutch engineering firm that had agreed to design the living quarters for workers in a Middle East dictatorship. During the design process, it became clear that there were some rather peculiar requirements that led the engineers to an unavoidable conclusion: this building was going to be a prison! The country's reputation did not suggest that the engineers were contributing to the fight against crime. Should they just do their design job, within the constraints of the programme

of requirements? The engineers were unwilling to complete their design.

In designing, sustainable development is often regarded as just an extra demand that can be added to the programme of requirements along with various performance and economic characteristics.

> Suppose that sustainability was an extra design criterion for a coal-fired power plant: where would it lead us? Well, the designer would have a huge problem!

The reason is that sustainability is not an add-on criterion. It is about all characteristics that a design should meet. It can even pose the question: should the design have been commissioned at all?

The commissioner and end-user of the product are frequently considered to be responsible for its sustainability. The contribution of the designer is regarded as relatively small and bounded by the programme of requirements for the process or product to be designed. Often this does not entail any more than selecting environmentally sound materials or production methods—choices that concern only the environmental part of the concept of sustainable development. Should designers want to make a real contribution to sustainable development, they will have to enter into discussions with the commissioner and the user.

The design process

Most engineers concern themselves with physical products in the broadest sense, i.e. every physical system designed with a certain purpose is a product. A bicycle is a product, but so is a factory, a water treatment facility, a bridge and a new area of a city. All these are physical systems, albeit it on different scales. Designing is the developing and planning of such a product.

Different disciplines have different design processes, though they share some common characteristics. The design process can be characterised by the cycle of design,[1] which describes certain steps present in each design process (Figure 8.1). When designing for sustainable development, designers should bear it in mind throughout the design process.

The most important decisions concerning sustainable development take place in the initial phase of the design process—analysis; the earlier sustainable development plays a role in the process, the larger its influence. It is much easier to alter the assignment to improve sustainability than trying to increase the sustainability of an already finished detailed design.

---

1 N.F.M Roozenburg and J. Eekels, *Productontwerpen, structuur en methoden* (Utrecht, the Netherlands: Lemma, 1998).

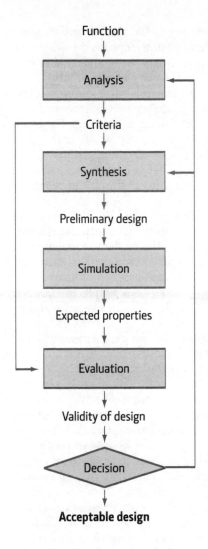

FIGURE 8.1 **The cycle of designing**

Analysis

The core of the design process is the function of the product that is to be designed. 'Function' does not only mean the technical function, but also any social, cultural, psychological and economic functions the product will perform. Although the programme of requirements will contain the main functions, these need to be analysed for further requirements.

Every design process follows from a problem that needs to be solved. It is the designer's job to identify the actual problem, clarify it and express it in a problem statement, i.e. the designer analyses what the real problem is. This often turns out to be a different problem from the one expressed by the commissioner. Only when the problem is defined clearly and in 'do-able' terms can the designer search for the most sustainable way of solving it.

This point is illustrated by the following example. The commissioner wants improved packaging for a particular machine. This improvement is required because the machines sometimes get damaged in transit. The commissioner could describe the problem as 'bad packaging'. If the problem is not defined as this but alternatively as 'damage of machine in transit', the designer may perhaps reach another solution that is more sustainable. It may be possible to change the design of the machine so that it can be transported safely without any packaging at all.

A group of people are usually involved in the problem (the 'problem owners'). First, there is the commissioner, who is not satisfied with the current situation. This could be because the profit from the product is too low, an inefficient production process, a market share that is too small or new legislation or technology demands a change in future.

Other problem owners may include end-users, sales people and people who live or work in the vicinity of the product (see the discussion on stakeholder management in Chapter 7).

The designer has to determine which aspects of the product are not sustainable (enough) and which can be improved on.

Once the design problem is determined, the design goal can be formulated. This goal describes the viable future situation that the design should realise. Determining the viable situation is in itself a problem-solving process and entails choices. What does this viable situation look like? There is no one correct, sustainable answer to this question. Sustainability leaves scope for choice.

The designer may conclude that major changes need to be made for a sustainable solution. But the commissioner will frequently not allow the designer to bring about such a large change. The programme of requirements is generally so restrictive that only small product improvements are possible. The engineer should therefore review the commissioner's programme of requirements critically before accepting a design assignment.

This design goal is subsequently transformed into functional and technical requirements. The functional and technical requirements fulfil different roles during the design process:

- They force all parties involved to reach a consensus regarding the product to be developed in an early stage of its development. This prevents the design from having to be overhauled at a later stage if one party regards the product as not living up to expectations

- It is a way of directing the design process. Preliminary proposals are tested against the functional and technical requirements

- It limits the area in which possible solutions can be found

- In some cases, it forms a part of the contract between the commissioner and the designer. The designer then has to deliver a design that adheres to the functional and technical requirements

Synthesis, simulation, evaluation, iterations

The actual designing takes place in the synthesis phase when the designer generates a possible solution to the problem described in the analysis phase. Generally multiple (partial) solutions are devised during the synthesis phase, of which one is finally chosen. The different solutions are described in writing, drawings, mock-ups, etc.

In the simulation phase, the designer tries to obtain an impression of the behaviour and characteristics of the designed product. This can be done through calculations, testing the performance of a physical model, computer simulations, etc. These simulations yield information about the expected behaviour of the product.

The preliminary design is tested in the evaluation phase. Here the characteristics expected from the simulation results are compared with the characteristics that follow from the functional and technical requirements. If the differences are acceptable, designing can continue into the next phase; otherwise adaptations are required.

These steps do not always happen in the same order. Designing is an iterative process and a previous step often has to be revisited. The final design is often reached through trial and error.

For large system designs, the evaluation is often a painstaking effort. As well as testing the design of all the elements, this phase should involve testing dynamic system characteristics as such how to start up a chemical plant and how emergency shutdown should be implemented. This often involves mathematical modelling. The risk analysis of a systems design should be repeated to prevent any hazardous course of events from being overlooked.

Social analysis

Stakeholder analysis and trend analysis in the design process

Sustainable development is not the consequence of the actions of one person or one organisation. It follows from joint actions and the pursuit of common goals. Design for sustainable development should aim therefore to involve all people with a stake in the realisation of the product or process (i.e. the stakeholders; see Chapter 7). However, the product or process may be of importance in one of the competitive arenas in our society. For example, market competition prohibits transparency to competitors and therefore also to the public. National and political competition and threats of terrorism prohibit transparency regarding some technologies.

Instead of reaching consensus with stakeholders, one could analyse their motives and arguments. A stakeholder analysis shows you systematically which parties play a role in the emergence and the solution of a problem. A trend analysis tries to reveal changes in society that are relevant to products and markets, e.g. an ageing population or more people living alone. As the design process is about future products, it is important to extrapolate any observed trends into the future.

By considering all stakeholders, a design can be made that best fits their needs and wants, and the trends observed in society.

Categorising stakeholders

The first step in a stakeholder analysis is to list all stakeholders and their characteristics. They can then be categorised according to a matrix depending on their influence and awareness (Figure 8.2). Four different categories of stakeholders are apparent in Figure 8.2).

Dividing the stakeholders into groups showing different levels of interest and awareness highlights why a certain situation is happening or where people stand regarding it. In addition, the most influential stakeholders can be identified.

There are four groups of stakeholders (Figure 8.2). **'Key stakeholders or key players'** (D) are strongly aware of their interest in the problem and have a lot of influence in the development of a solution. When solving the problem, this group of people is most important in discussing solutions. On the opposite side of the matrix are the **'minimum effort'** (A) stakeholders. These stakeholders have little awareness of their interest in the problem and have little influence on the solution. A stakeholder analysis may show, for instance, that too much attention is being given to these stakeholders.

The third group of stakeholders is the **'create awareness'** group (C). These stakeholders have the potential and the means to influence the situation, but are not (yet) inclined to exert influence on this problem. The **'give voice and influence'** group (B) has little influence (yet) on the situation, but knows it has an interest in further developments. This group should be kept up to date on the latest developments and, where possible, obtain influence in developing the situation. This group of stakeholders could enhance its influence over time.

Trend analysis

A trend analysis tries to reveal changes in society that are relevant to products and markets. A trend analysis can complete the stakeholder analysis, especially in problems concerning sustainable development. This trend analysis can be performed in two ways:

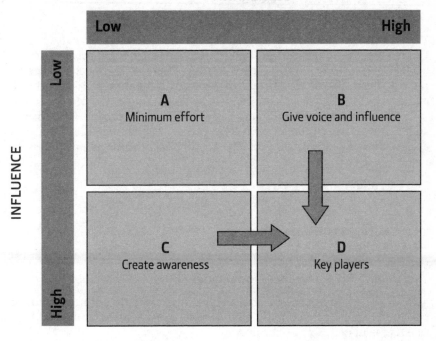

AWARENESS

	Low	High
Low	**A** Minimum effort	**B** Give voice and influence
High	**C** Create awareness	**D** Key players

FIGURE 8.2 **Influence–awareness portfolio**

- Determining trends per stakeholder
- Determining global trends and identifying the effects on stakeholders

The advantage of the first method is that trends may be found that will not show up in a global analysis. In addition, this method will identify which stakeholders can influence a trend and to what extent they can do so.

The advantage of the second method is that new stakeholders can be found that are not evident in the current situation, e.g. the future residents of a neighbourhood that is being designed and the next generation of consumers. Moreover, global trends may influence the relative importance of stakeholder groups, which is difficult to observe using the first method.

Trends can be analysed and combined. In complex situations, where there are mutually exclusive options, scenario analysis might help (see Chapter 10).

Life-cycle analysis (LCA)

Method

LCA is a tool that allows the total environmental impact of a design or a product to be analysed. It can be used during different phases of the design process. It can also be used to optimise the environmental performance of a design.

LCA quantifies the environmental impact of a certain product-system. The LCA of an existing product or system can set the bottom line for a new design. The product system encompasses all phases of the product life, i.e.

- Raw materials acquisition and refining (e.g. mining, drilling, agriculture, forestry, fisheries)

- Processing and production of product and production equipment

- Distribution and transport

- Use, re-use and maintenance

- End-of-life—landfilling, incineration, litter and recycling[2]

In all these phases, the contribution of the product to different forms of pollution (e.g. the greenhouse effect, ozone layer depletion and acidification) is calculated.

In LCA, all the contributions of a product to a specific environmental problem are counted in the same standard unit. For example, all greenhouse gas emissions are recalculated to their carbon dioxide (CO_2) equivalents in kg and all acidification is recalculated to the equivalent weight in kg of sulphur dioxide (SO_2).

However, these different forms of environmental impact cannot be added together. In order to calculate one single number as the result of the LCA, weight factors have to be introduced that set the relative priority for each environmental problem. Weight factors can be derived from the relative distance of the current situation in regard to the goals set out in policy documents.[3]

Alternative designs and materials can thus be compared. However, it is important to remember that the single number is always relative to the choice of weight factors. If priorities change, the LCA score will change too.

2 R.M. Bras-Klapwijk, R. Heijungs and P. van Mourik, *Levenscyclusanalyse voor onderzoekers, ontwerpers en beleidsmakers* (Delft, the Netherlands: VSSD, 2003).

3 In principle, the method gives a higher weight to those environmental effects for which policy targets are most ambitious. The precise method is given in: R. Heijungs, J.B. Guinee, G. Huppes, R.M. Lankreijer, H.A. Udo de Haes, A. Wegener Sleeswijk, A.M.M. Ansems, P.G. Eggels, R. van Duin and H.P. de Goede, *Environmental Life Cycle Assessment of Products*. I. *Guide*. II. *Backgrounds*. (Leiden, the Netherlands: CML, 1992).

The subject of the analysis, the product or process, is expressed as a so-called functional unit. Setting this unit is very important because it determines the subject of the analysis and what the results can be compared to.

One example of a choice for a functional unit is a car of specific characteristics. In this case, the total environmental impact of the car during its entire life-cycle is calculated. Different cars can then be compared with each other.

Another example of a functional unit is 'the transport of two adults and two children, carrying a total of 60 kg of luggage from Amsterdam to Barcelona'. In this way, different candidates for transport can be compared, e.g. car, aeroplane, bus and train.[4]

LCA can also be used if a product is being redesigned, when the baseline design can be analysed to show main points of concern for environmental impact. This can help in directing the search for improvements.

Not many data are available at the start of a design process and a global analysis can be carried out only to compare various existing alternatives with respect to environmental impact. Subsequently, LCA can help in evaluating the design and provide clues for improving it.

The LCA forces the designer to view the entire product life-cycle. This ensures that important environmental problems are not overlooked. The catalytic car exhaust example (see over) shows how important the two main functions (determining environmental impact and comparing different products) of an LCA can be.

Tools

A number of software tools can be used to prepare an LCA. These vary from those that can be used to look up the environmental damage of materials (e.g. IdeMat[5]) to those that help the designer to make a short (e.g. EcoScan®[6]) or detailed (e.g. SimaPro[7]) environmental analysis.

One manual method that has been developed for product designers is the Eco-indicator-99 method.[8] This method gives a table filled with quantities of used materials and processes for the production, utilisation and disposal phases of the product. The listed values are multiplied using standard eco-

---

4 The results are strongly dependent on the load factor of each mode of transport (actual passengers/maximum passenger capacity); the bus is clearly the winner. J.H.J. Roos, A.N. Bleijenberg and W.J. Dijkstra, *Energiegebruik en emissies van de luchtvaart en andere wijzen van personenverkeer op Europese afstanden* (Delft, the Netherlands: CE Delft, 1997).

5 www.io.tudelft.nl/research/dfs/idemat/index.htm, 11 November 2005.

6 www.ind.tno.nl/en/product/ecoscan, 11 November 2005.

7 www.pre.nl/simapro/default.htm, 11 November 2005.

8 www.pre.nl/eco-indicator99/default.htm, 11 November 2005. More details of the method are given in: Netherlands Ministry of Housing, Spatial Planning and the Environment (VROM), *Eco-indicator 99 Manual for Designers* (The Hague: VROM, 2000); www.pre.nl/download/EI99_Manual.pdf, 11 November 2005).

Catalytic car exhaust

A car exhaust catalyst performs admirably in terms of its environmental impact if it is regarded only in its usage phase. The release of exhaust products such as carbon monoxide (CO), volatile organic compounds (VOCs) and nitrogen oxides (NO_x) has fallen dramatically since catalytic exhausts were made mandatory for cars in the EU in 1993. Because this technology required the introduction of unleaded fuel, this also put an end to lead pollution from this source. The car exhaust catalyst can thus be seen as an environmentally sound product.

But if the entire life-cycle of the catalyst is considered, other conclusions can be drawn. The precious metals platinum, palladium or rhodium are needed to produce a catalyst. Production of every three-way car exhaust catalyst requires 4.5 grams of platinum and 15 grams of palladium. For every gram of these metals, 300 kg of ore is mined.

The market leader in precious metals is situated in the Siberian industrial town of Norilsk. The construction of the factory and mining complex began 70 years ago. Since then, no environmental measures were taken. This complex is said to be Russia's—and possibly the world's—largest polluter. Annual emissions of SO_2 are estimated at 2.8 million tons—about as much as Germany's entire emissions and 20 times those of Sweden.

A permanent blanket of SO_2, NO_x, CO, phenol and chlorine covers the area. The town itself is devoid of trees and at least 4,000 km² of forest is affected. The average life expectancy of the male inhabitants of Norilsk is low (49 years). Thus, a number of things could be improved if we consider the total LCA of the car exhaust catalyst.

The efficiency of a car engine is strongly related to its compression ratio. In the 1980s when electronic injection systems were becoming commonplace, values around 1:9.5 became normal with octane 98 leaded petrol. For proper functioning, however, car exhaust catalysts require 95 octane unleaded petrol. With the introduction of the new petrol, the compression ratio had to be reduced to 1:8, which caused fuel consumption to rise.

Hence, a choice was made between using less fuel per kilometre but emitting a higher percentage of toxic exhaust products (i.e. engine without catalyst) and a higher level of fuel consumption but 'cleaner' exhaust products (i.e. with catalyst). The latter option was chosen, though this choice is not obvious as the consumption of fossil fuels and the related emissions of carbon dioxide are higher.

Source: N.P. Walsh, 'Hell on earth', *The Guardian*, 18 April 2003; www.guardian.co.uk/g2/story/0,3604,939043,00.html, 11 November 2005); and J. Trommelen, 'Schone auto vervuilt Siberië', *DeVolkskrant*, 18 January 2003

indicators. This allows the designer to determine which processes, components or phases of the life-cycle of a product contribute most to its environmental burden. Several product alternatives can also be compared.

Figure 8.3 shows the input factors for a coffee machine. The following assumptions are made:

● Lifetime of five years

● Used twice per day at half capacity

● The coffee is kept hot for 30 minutes

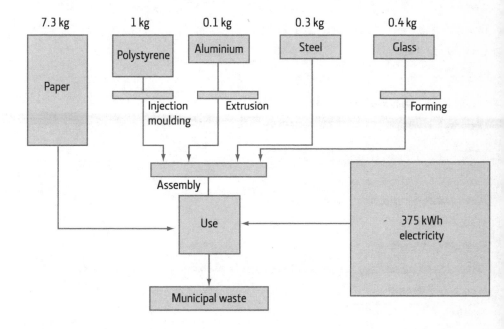

FIGURE 8.3 Input factors for a coffee machine

Source: www.pre.nl, 11 November 2005; www.pre.nl/eco%2Dindicator99/eco%2Dindicator%5Fapplication.htm

Table 8.1 shows the completed eco-indicator table for the machine.

LCA is an effective way of comparing alternatives in the design process, especially when various environmental aspects play a role. It also can be used to determine the priorities for environmental improvement.

However, LCA has its drawbacks and problems. It sometimes takes an excessive amount of time in comparison with the gains to be achieved. A quick but rough scan is generally sufficient to eliminate the main alternatives.

Material of process	Amount	Indicator	Result
Production: materials, treatment, transport and extra energy			
Polystyrene (PS)	1 kg	360	360
Injection moulding of PS	1 kg	21	21
Aluminium	0.1 kg	780	78
Extrusion of aluminium	0.1 kg	72	7
Steel	0.3 kg	86	26
Glass	0.4 kg	58	23
Gas-fired heat (forming)	4 MJ	5.3	21
Total [mPt]			536
Use: transport, energy and possible auxiliary materials			
Electricity (low voltage)	375 kW	37	13,875
Paper	7.3 kg	96	701
Total [mPt]			14,576
Disposal: materials, treatment, transport and extra energy			
Municipal waste, PS	1 kg	2	2
Municipal waste, ferrous	0.4 kg	−5.9	−2.4
Household waste, glass	0.4 kg	−6.9	−2.8
Municipal waste, paper	7.3 kg	0.71	5.2
Total [mPt]			2
Total all phases [mPt]			15,114

mPt = milliPoint (the Point is the unit used by the Eco-indicator 99 method)

TABLE 8.1 **Completed eco-indicator tables of a coffee machine**

Source: www.pre.nl, 11 November 2005

It is often difficult to determine the correct system boundaries. For example, in the case of recycling, are the costs involved in processing the recyclable waste related to the end-phase of the discarded product or to the acquisition of resources of a new product?

The method is less effective when toxic or dangerous substances are used but not emitted. It is based on a listing of input and output streams at a point in time; in the case of so-called 'product-encased toxicity', the toxic substance may leak from dumped waste decades after its disposal.

A similar problem arises in the case of discarded material disposed of to landfill. How should the emissions emanating from the landfill be taken into account?

LCAs are not applicable to every discipline. They are particularly useful for designers of products produced using 'off-the-shelf' methods and materials. These methods can be treated as static components in the analysis. Engineers designing completely new production processes or chemical processes have less use for an LCA.

Design tools and strategies

There is a large number of strategies that a designer can follow to design eco-efficiently. In the LiDS wheel (Lifecycle Design Strategies), these strategies are clustered and visualised (Figure 8.4). Each strategy contains a number of basic rules.

When the design process starts with an existing product that needs to be improved, the LiDS wheel can be used to create an environmental profile of this product. Each product can then be analysed according to the criteria of that strategy, using a measurement on a scale from 1 to 5. The analysis can be represented graphically by the spokes of the wheel (Figure 8.4). In this way, it becomes evident which strategies will yield the highest improvement.

The LiDS wheel is constructed in such a way, that moving clockwise, the strategies numbered 1 through 7 follow the life-cycle of the product from resource acquisition to waste disposal. Strategy number 0 ('optimising the way the function is performed') can be seen as a long-term improvement because it involves structural and radical solutions. These strategies are explained below.

Strategy 1: choose materials with low environmental impact

Chose materials that are:

- **Clean materials**. Choose non-toxic and harmless materials. For example, avoid materials containing heavy metals, asbestos, chlorofluorocarbons (CFCs) and endocrine disruptors such as phthalates

- **Renewable materials**. Avoid scarce materials, i.e. materials from a non-renewable or slowly renewing sources such as fossil fuels, copper, tin, zinc and platinum. Plastics too are made from fossil fuels and are counted as scarce materials

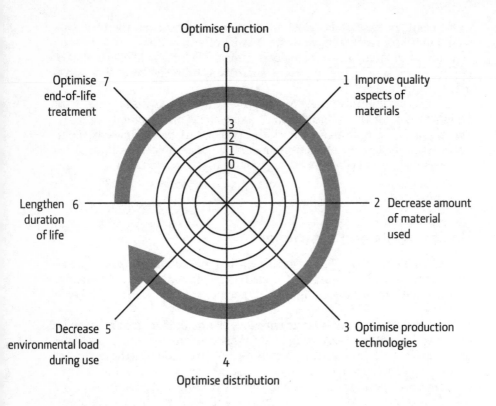

Optimise function
0

1 Improve quality aspects of materials

Optimise end-of-life treatment 7

2 Decrease amount of material used

Lengthen 6 duration of life

3 Optimise production technologies

Decrease 5 environmental load during use

4
Optimise distribution

FIGURE 8.4 **The LiDS wheel**

Source: C. van Hemel, *Ecodesign Empirically Explored: Design for Environment in Dutch SMEs* (Delft, the Netherlands: TU Delft, 1998)

- **Lower-energy-content materials**. Mining and production of some materials such as aluminium is very energy-intensive. However, the use of aluminium can have a positive effect as its light weight can reduce energy consumption by the product (e.g. a car). Aluminium should be recycled

- **Recycled materials**. These should be prescribed by the designer whenever possible. Metals have recycled metal added by default. Recycling of plastic products is currently limited and stimulating market demand will increase recycling

- **Recyclable materials**. Use materials that can be recycled. This will be even more effective if collection systems are in place or planned. Materials should be selected that will result in high-quality recycled materials. The fewer types of materials that are selected, the easier it will be to collect and recycle the materials

Strategy 2: dematerialise

The principles of dematerialisation are:

- **Reduction of weight.** Using less material often reduces the product's environmental impact. Particularly in transportation equipment, the gains can be enormous. Less material means less resource consumption, less waste and a lower environmental impact during transportation

- **Reduction in volume.** When the product and its packaging are reduced in size and volume, more products can be transported in a given transport facility, making transportation more efficient. Another solution is to make the product foldable or 'nest-able'

Figure 8.5 shows an example of dematerialisation.

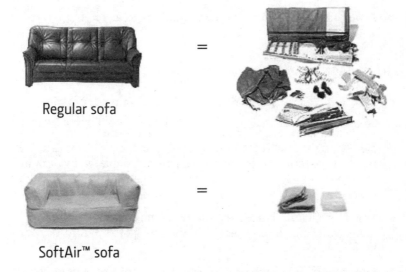

The SoftAir™ sofa is completely filled with air. The user inflates it at home using a hairdryer or a vacuum cleaner. Compared with a normal couch, this technology offers a 85–90% reduction in transport volume and a weight reduction of over 80%. The couch consists of 12–14 different compartments, which can be replaced individually should one of them be punctured.

FIGURE 8.5 **How to dematerialise a sofa**

Source: www.ikea.com, www.softair.com, 11 November 2005

Strategy 3: select environmentally efficient production techniques

- **Environmentally sound production processes**. For example, DSM-Gist in Delft replaced 15 chemical production steps in pharmaceutical production with five biotechnological ones. This drastically reduced the consumption of water and harmful emissions

- **Fewer production steps**. For example, combining various parts into one cast item can save assembly steps and simplify the design. Chose materials that require no finishing touches or only apply finishing to parts that absolutely require it (e.g. because they are visible)

- **Lower/cleaner energy consumption.** Make efficient use of (sustainable) energy sources. Use sustainable energy sources such as windmills and solar cells in the production process. Also use waste heat from other companies situated nearby

- **Reduce production waste**. Strive for minimum waste in production. Prevent waste by designing the product so that its shape resembles the shape of the base material

- **Fewer/cleaner production consumables**, e.g. cooling water, degreasing agents and paper

Strategy 4: select an environmentally sound distribution system

- **Less and cleaner packaging**. Use less packaging material or packaging with a lower environmental impact, e.g. replace polystyrene packaging with cardboard. Bioplastics made from starch, for example, can be a good replacement for plastic foams in some applications. A reduction in packaging can also be obtained by designing the product in such a way that there is no need for packaging or less packaging is needed

- **Energy-efficient transport mode**. Some modes of transport are more polluting than others. Freight transport by train or boat is preferable to transport by road or plane

- **Energy-efficient logistics**. The cleanest transport is no transport. Nowadays, production sites in a supply chain are often far apart. Reducing transport distances can result in significant energy savings

Strategy 5: reduce environmental impact in the use phase

A product also has an environmental impact during the use phase. This should be considered in the design process. The following principles hold:

- **Reduce energy consumption during use.** For example, energy consumption by a vehicle is generally far greater than the energy consumed in its production

- **Cleaner energy sources**. Enhance the use of renewable energy sources such as solar energy (Figure 8.6), wind energy, tidal energy and energy from biomass. With some products, even human power can be used. Various watches use the movement of the wrist of the wearer to charge a capacitor, which drives the mechanism. A number of wind-up products are being introduced, e.g. wind-up flashlights, radios, navigation systems and mobile phone chargers

- **Fewer and cleaner consumables needed**. Make sparing use of consumables such as cleaning agents, maintenance materials, lubricants, water, light bulbs, cartridges and coffee filters

Strategy 6: optimise the life-span

A product with a long life-span will have a lower environmental impact because fewer materials are used overall. Prolongation of the technical life-span can be obtained by making:

- The product more reliable and easier to repair

- Maintenance easier

With a modular system, only a small part of the product needs to be replaced when it fails, instead of the entire product. Enhancing the reliability sometimes means the use of more materials, e.g. making the construction more robust.

An elongated life-span is useful only if the duration of usage is also prolonged. Products that are highly fashion-sensitive (e.g. clothing and mobile phones) are often discarded long before their technical life-span has expired. Fashion sensitivity fuels consumption. For example, the success of SWATCH[9] is mainly because it succeeded in making its watches trendy in order to make customers buy several watches each year.

Enhancing the relation between user and product may also prolong use. Modularity can help too; by merely changing one part of the product (e.g. a telephone cover), the product can keep up with fashion and thus be used longer. If it is not possible to prolong of the duration of use, one alternative is to shorten the technical life-span in order to save on materials. In short, the rules of thumb are:

- Reliability and durability

- Easier maintenance and repair

9   www.swatch.com, 11 November 2005.

Figure 8.6 **Solar boat at Nieuwkoop, Netherlands**

- Modular product structure
- Classic design
- Stronger product–user relation

Strategy 7: optimise end-of-life system

- **Stimulate re-use of the entire product**
- **Remanufacturing/refurbishing**. Stimulate re-use of parts. But to enable the re-use of parts, it is necessary that the product can be dismantled easily. This means, for instance, avoiding glued connections
- **Recycling of materials**—the less variation in materials used, the more efficiently materials can be recycled
- **Safe incineration**. Avoid the use of substances that generate hazardous fumes when incinerated

Strategy 0: optimise the way the function is performed

Is the product the most sustainable way of providing a service to a user? Sometimes new solutions can be developed. This strategy does not aim to improve an existing product, but to design new ways of fulfilling a need. A number of principles can be applied:

- **Dematerialisation**. This is different from strategy 2 ('reduce material use'). The crux is to replace a material product by an immaterial way of fulfilling the function. The material product is usually replaced by a service or a product–service combination

- **Shared use of the product**. Design a product to be used by multiple users. The users will use the product more efficiently through shared use

- **Integration of functions**. The integration of different functions into one single product saves materials and space, e.g. a television with a built-in video recorder or a combination of a fax/printer/scanner. This approach is only useful if the individual components can be repaired separately in order not to diminish the durability of the product

- **Functional optimisation of products (components)**. If a critical look is taken at the way a product fulfils its function, some parts may turn out to be unnecessary. This applies to products where an impression of quality or status is realised through the use of extra or oversized parts

- **Create awareness with respect to the quality and origins of a product**. Most people know little about the products they use. What material is it made of? Where was it manufactured? And under what circumstances? To stimulate consumers to switch to sustainable consumption, it is important that people gain a better understanding about the origins of products, their effects on the ecosystem and the social conditions at the manufacturing site. Ideally, the product itself should reveal this information

Questions, discussion and exercises

1. Download an LCA demo programme and explore its limitations. Examples include:

   www.pre.nl/simapro/download_simapro.htm, 11 November 2005

   www.environmental-expert.com/software/pr_eng/form.htm, 11 November 2005

2. The environmental performance of a product depends strongly on the way it is used (e.g. the number of passengers in a car). The use of a ceramic coffee mug in preference to a disposable polystyrene coffee cup has been the subject of considerable discussion. Which usage characteristics would be important in this choice?

3. How could you strengthen the bond between product and its owner? How is this important for sustainable development?

9 Innovation processes

The design process is often not the way in which a new product or process is delivered. In reality, many new technologies are not commissioned but are frequently the result of painstaking efforts by scientists and engineers, which are sometimes even ridiculed by their employers. Success is often unexpected as the new technology becomes successful due to the new and unforeseen options it provides which enable customers to behave differently.

Technological innovation is more than just design. It is often the result of 'heterogeneous engineering' involving a wide array of activities. Innovation is frequently fuelled by the enthusiasm of scientists and engineers, and steered by paradigms that guide knowledge creation and objectives for improvement. Of course, financial considerations play a role. But the artistic pleasure of creating a new machine that really works is often underestimated.

The previous chapter stressed the importance of stakeholder analysis before starting a design and, in Chapter 5, we discussed briefly the historic relationship between technology and society. Building on these themes, this chapter explains the main concepts regarding the dynamics of technological change and shows why technologies tend to be resistant to social demands to change them. In Chapter 10, we will reflect on the question of how to create really sustainable technologies and how interfacing points can be used to influence technology towards sustainability.

The demise of the linear model

Historically, science has always been a 'cultural' activity financed by government, private benefactors or non-commercial organisations. Innovation was a responsibility of industry. As science and technology became more related and governments recognised the value of science and innovation for the national economy, policy questions emerged regarding the role and level of government funding for research (e.g. research organisations and universities). Models were developed regarding science, technology and their relevance for the economy.

The linear model assumes a linear sequence that starts with basic research and leads to the use of new products by consumers (Figure 9.1).

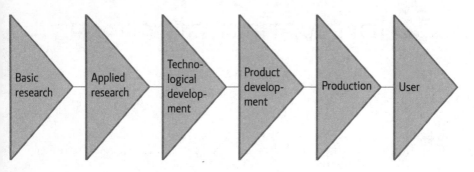

FIGURE 9.1 Six-phase linear model of an innovation process

- This model supposes that technological innovation begins with the performance of basic scientific research. This leads to certain results deemed to be of commercial interest, e.g. a new plastic material with high electric conductivity

- These scientific results are elaborated by applied research. What products could be made using this new material? What will be the basic outline of the process to make it? How much is it likely to cost to produce it?

- During the technological development phase, the first pilot production plants are developed, tested and evaluated. Based on the results obtained, the actual production processes for the material are designed and then for any auxiliary chemicals needed

- In product development, the material is tuned to other market demands. In what shapes, colours, and quantities are the materials needed? What grades of quality are needed? Is it necessary to supply customers with additional products so that they can use the product?

- Production (manufacture, distribution and marketing) just has to pick the fruits of earlier phase, i.e. carry out the planned processes and deliver the product

- If all these phases are satisfactory, the user accepts the new product gratefully and everyone is happy

Criticism of such models began in the 1970s and focused on a number of points:

- The innovation process is far more complex than depicted in linear models. Choices are made continuously between alternative development options. Social factors play an important role in such choices. Feedback loops are important and previous choices often have to be reconsidered

- New technologies do not necessarily start with basic research. Ideas may arise at various stages and basic research may be influenced by user ideas

- The time dimension is neglected. Even if all the actors could be brought together in a model that allowed more complex interactions, the time factor has to be added. Applied researchers and process/product developers learn gradually how to use the various possibilities offered by new scientific theories. Users adapt gradually to the changes that are introduced by new technologies. Their use of the product is often dependent on the creation of infrastructure, which is again dependent on user interest. Scientists recognise new demands for knowledge only gradually. Time lags lead to a dynamic pattern in the introduction of new technology

- The linear model is an example of a **'black box'** approach. No attention is given to the qualitative choices on technologies made within the project, i.e. they remain within the black box

It became clear that the development of new technology is not a linear process determined only by scientists and technologists. Students of technological innovation therefore tried to form a more comprehensive framework encompassing innovations and various stakeholders where the influence that society has on the creation of technology could be addressed.

An important stimulus came from the history of technology, which is based on both internal and external approaches:

- The **internal** approach is concerned with the precise tracking of the origins of the characteristics of specific machines, e.g. steam engines. For example: who developed a specific kind of valve for the steam engine? What were the scientific origins of the idea? How did this valve spread in the steam engine world?

- The **external** approach focuses on the social environment of a technology. In this approach, a new valve on a steam engine would be described as an answer to problems that steam engines created for people such as their operators, owners or producers

The internal approach generally leads to studies on changes in techno-scientific knowledge. The geniuses that introduce new concepts are the heroes of the story. The external approach creates fewer individual heroes, but shows the links between technology and processes in society.

In the next section, we examine the SCOT model—introduced in Chapter 5—in more detail, using the development of the bicycle as an example. We then consider another model for technological development—the so-called 'system model' and discuss economic approaches to innovation.

Social construction of technology

A more recent approach is the SCOT model (Social Construction Of Technology). In this model, technologies are considered to be social constructions to which various groups of people have given shape. One of the goals of the SCOT model is to indicate that various choices are being made. Social constructivism means that all techniques are to be regarded as human-made objects, which reflect the choices of humankind; techniques are *not* the imperfect reflection of an ideal object that has come to us by science or divine revelation.[1]

An analysis of historical development of the bicycle will help us to explain the SCOT model. If we consider the development of the bicycle in the traditional linear way, the current safety bicycle (Lawson's Bicyclette) is seen as the final product of an evolution which started with the 'walking machine' or 'boneshaker' and where the Penny-Farthing represents a transitional point (Figure 9.2).

The problem with such a representation is that it is not clear that choices have been made. What seems odd is that the Penny-Farthing stayed in the picture for so long and even when the safety bicycle had been introduced, even though it was technically inferior.

The Penny-Farthing has a large front wheel and a small rear wheel. Pedals attached to the forward axis propelled the bicycle. The need to reach the handlebars and steer the bicycle meant that the cyclist had to sit almost directly above the front wheel. The bicycle was fast and efficient, but highly unstable. It was introduced in 1870 and was popular until the end of the 19th century.

Lawson's Bicyclette was characterised by rear-wheel transmission. This bicycle stems from 1879 and, by the end of the 19th century, its design had crystallised into the safety bicycle as we know it today. Other typical characteristics included wheels of the same size and air-inflated tyres, which contributed greatly to its safety.

1   W.E. Bijker, T.P. Hughes and T. Pinch (eds.), *The Social Construction of Technological Systems: New Directions in the Sociology and History of Technology* (Cambridge, MA: MIT Press, 1987).

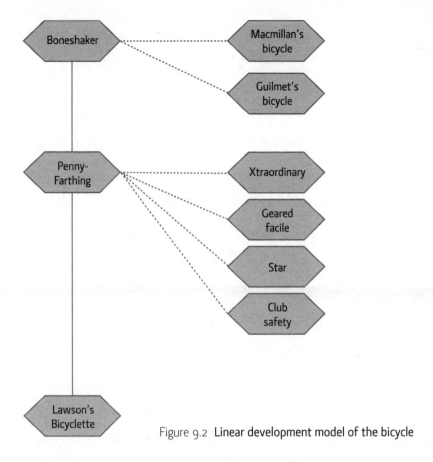

Figure 9.2 **Linear development model of the bicycle**

To better understand the development of the bicycle, we need to appreciate two important concepts in the SCOT model, i.e.

- **Artefact**: a consciously human-made, artificial object
- **Relevant social group**: people who are involved in a certain technical development and who hold the same view regarding that artefact

Around each artefact, a number of relevant social groups can be distinguished (Figure 9.3). People involved with an artefact all hold a certain image of it, i.e. they attribute a certain meaning to it. Especially important is what people regard as problematic about the artefact.

Groups can have different problems with regard to the artefact (Figure 9.4). For problems (or clusters of problems), multiple solutions are conceivable (Figure 9.5). The different social groups exert influence on the develop-

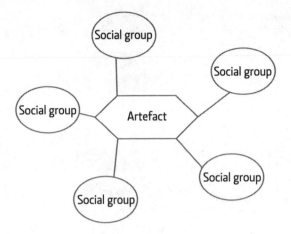

FIGURE 9.3 **Artefacts perceived by social groups**

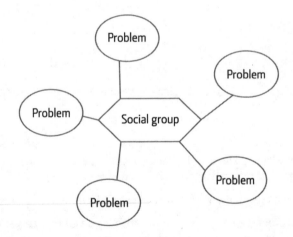

FIGURE 9.4 **Social groups defining problems**

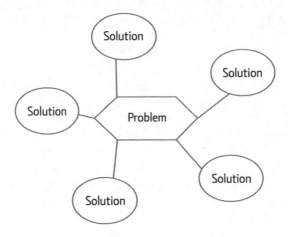

Figure 9.5 Social groups selecting solutions

ment through their way of defining the problem and possibly including their solution(s).

Using the cluster 'technical artefact–relevant social group–problem–solution', we can explain the development process of an artefact. Flexibility of meaning is important here as different social groups attribute different meanings to the same technical artefact. This notion is essential in understanding, for instance, the development of the bicycle.

Relevant social groups surrounding the artefact 'bicycle' include the producers and the users of the bicycle. However, we should also involve the 'anti-cyclists' in the story as the bicycle also met with resistance:

> . . . but when to words are added deeds, and stones are thrown,
> sticks thrust into wheels, or caps hurled into the machinery,
> the picture has a different aspect.[2]

These anti-cyclists were concerned about the 'decency' of female cyclists and argued against bicycles because of the dangers involved. In London, for instance, cyclists used wooden pavements because the streets were not paved. This upset local people. The conflict was further enhanced by class differences. The largest group using the Penny-Farthing turned out to be young men of reasonable wealth who possessed the courage and dexterity to handle the machines. These Penny-Farthing riders radiated superiority towards others who were walking or riding horses. To them, the Penny-Far-

2 W.E. Bijker, *The Social Construction of Technology* (Eijsden, the Netherlands: Proefschrift Universiteit Twente, 1990): 47.

thing was a 'macho machine'. Potential users such as women, long-distance cyclists or older men tended to consider the Penny-Farthing as an unsafe machine.[3]

The different attributions of meaning also spawned different directions of development:

- For the sporty Penny-Farthing users, the best way to increase speed was to enlarge the front wheel. This culminated in 1892 with the Rudge Ordinary, which had a front wheel of about 1.4 m in diameter. The fact that this made the bicycle even more dangerous was considered by its users to be more of an advantage than a drawback

- To make the unsafe Penny-Farthing suitable for other users, experiments were carried out with various different models. The wheels were reversed (e.g. with the Pony Star) or the front wheel was made smaller and the saddle was put further backwards as in Lawson's Bicyclette—in which we recognise the modern bicycle

Lawson's Bicyclette was not an immediate success and, for decades, it was only one of a number of alternatives.

Its final success was not due to its characteristics as such. After the development of inflatable or pneumatic tyres by John Boyd Dunlop in 1888, the ride of all bicycles improved considerably. With Lawson's Bicyclette, speed could be increased by adapting the transmission. But the same trick could not be applied to the design of the Penny-Farthing. Lawson's Bicyclette therefore developed into a faster bicycle than the Penny-Farthing. This killed the 'macho' image of the Penny-Farthing and its fate was sealed.

Figure 9.6 shows that the application of the SCOT model results in a different view of the development of the bicycle from a traditional phase model. In the SCOT model, the assignment of meaning, the signalling of problems and solving them are aspects of the process of technological development. It is entirely possible that the final product was not a result foreseen by any of the parties involved.

This is certainly true for the safety bicycle, whose development cannot be clearly dated but which stretches over a number of decades. At the start of the process that led to this bicycle, none of those involved had a clear vision of it in its final form. Several different models were also present at that time, including Lawson's Bicyclette, the Kangaroo and the Facile.

This example illustrates how the development of an artefact consists of three (repeating) stages:

---

3 Because the cyclist was seated almost directly over the middle of the front wheel with their legs far above the ground, every stop or bump brought the risk of falling off.

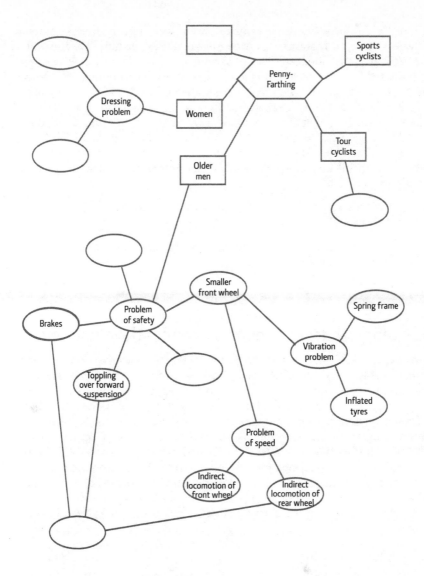

FIGURE 9.6 **SCOT model applied to the bicycle**

1. Interaction between technological development and social groups

2. Variance in problems and solutions

3. The choice for one solution

Of course, the solution does not need to be permanent; the solutions obtained are temporary forms of stabilisation. And what is used as a solution does not have to be an example of brilliant engineering. In the case of the Penny-Farthing, for instance, users complained that the handlebars would be in the way if the rider fell forward as a result of a sudden stop. The solution proposed for this problem was to make the handlebars detachable, thus allowing riders to land on their feet. Although such example solutions could not be dismissed outright, the solution did not last long in this case.

The SCOT model is not the only model that primarily attributes technology to the social world. What the SCOT models lacks is a good analysis of the interactions between relevant social groups, who do not have an equal opportunity to influence the process of technological change. Influence depends on:

- The resources of the various actors (money, know-how, status, organisational position)
- The pattern of the social relations in which they are engaged, i.e. social network formation

Technological development as system development

Individual artefacts play the key role in the social-constructivist and the evolutionary–economic views of technological development. Thomas Hughes, a US technology historian, in 1883 posed another, opposing vision, i.e. that it is not so much the development of individual artefacts that matters, but that the formation and further development of technological systems is crucial. Though using an obvious example to illustrate his view (electrical systems— see below), Hughes held a broader view. He believed that the basis of the system-wide approach to technological development is the assumption that the development of all large-scale technology (not only electrical systems) can be studied as a history of developing systems. The history of a specific system, i.e. the electrical system, thus serves only to illustrate a more general view.

Technological systems[4]

The basic assumption when analysing technological change from a systems perspective is that a problem in one part of the system may lead to a technological change somewhere else in the system. In the systems approach, the problems that were attributed to the bicycle could also lead to changes in the production of bicycle materials such as steel pipes for its frame, or to changes in road construction. These elements can be considered part of the bicycle system.

4   T.P. Hughes, *Networks of Power: Electrification in Western Society 1880–1930* (Baltimore, MD: Baltimore University Press, 1983).

A system consists of components or parts that are related to each other. These components are connected by a network or structure, which may be more interesting than the components themselves. The linked components of a system are often controlled centrally and the system boundaries follow from the range of this central control.

Control is present to optimise the performance of the system as a whole and to direct the system towards a common goal. The goal of an electricity production system, for instance, is to convert the available sources of primary energy (the input) into electricity (the required output) and thus fulfil the demand for an easy accessible source of power. Because the components are linked to each other, the state of one component influences that of another.

Elements that do not belong to the system but have an influence on it are collectively called the 'environment' of the system. Parts of the environment can sometimes be made part of the system by bringing them under the system's control.

An open system is subject to environmental influences; a closed system is not. Hence, the behaviour of a closed system is, in principle, completely predictable—that of an open system is not.

Reverse salients and critical problems

The growth of systems is analysed through the concepts known as:

- Reverse salient
- Critical problem

Because system components often grow at different rates, parts of the system can be identified that lag behind the others in terms of growth and limit the growth of the system as a whole. The term '**reverse salient**' stems from military history. In that context, the term signifies a section of a front line that is lagging behind in the advance. This metaphor is well chosen because a progressing military battlefront often shows irregular and unpredictable behaviour—just as a developing technological system does. Military personnel frequently direct all their efforts towards fixing a 'reverse salient'. The same goes for the development of a technological system. Inventors, engineers, entrepreneurs and others direct their creative and constructive forces primarily at the correction of 'reverse salients' such that the system functions optimally in fulfilling its tasks.

If the reverse salients are identified, they will often be translated into a series of **critical problems**, i.e. problems that are well structured and that can, in principle, be solved. The redefinition of reverse salients as a series of critical problems is the essence of the creative technological innovation process in the system. An inventor, engineer or scientist transforms an

The electrical system

Systems that provide electricity consist of electricity generation, transformation, control, transmission networks, distribution networks and user components. Between 1880 and 1930, electricity was produced using steam engines, steam turbines and water turbines. Different types of generator types were coupled to these 'prime movers'. Transformers were very important for larger systems. User components included lamps, engines, heating equipment and electrochemical equipment.

The system served widely different purposes. Transmission distances during this period increased from a couple of hundred metres to cover a region of tens of thousands of square kilometres. Distribution networks transported the electricity from the transmission network to industries and homes. Control components regulated the electricity production system to maintain characteristics such as voltage and frequency at the right level, and to make sure the system functioned optimally regarding its goals such as efficiency, profit generation, reliability, etc.

The most difficult components to define are those at both ends—both on the demand and supply sides. For instance, are the mechanical 'prime movers' a part of the electricity system? Water power sometimes escapes control by the system. When you consider that the grid sometimes influences them (peak loads), are the different usage intensities parts of the system and sometimes not?

Thomas Hughes chose to take the 'prime movers' into the definition of the system because inventors, engineers and researchers treated them as such and because they were mainly controlled by the system.

amorphous challenge—the lagging-behind of (parts of) the system—into a series of problems, which are expected to be solvable.

This is a crucial and often underestimated part of the engineering profession, being able to redefine unstructured problems into a series of solvable, critical problems. The confidence in the solvability of the reverse salient increases dramatically once it has been turned into a series of well-posed critical problems. Correct articulation of problems usually helps considerably in approaching a solution. However, a situation sometimes occurs where a critical problem appears to be unsolvable.

Phases in system development

Different phases can be distinguished in Hughes's model for system development and different characteristics hold a dominant position in different phases of system development. Moreover, the model indicates the skills that managers must possess in each of the phases and the guiding interests.

The first electrical system: reverse salients and critical problems

Around 1880, Edison created the first electrical system in the world. This Manhattan electricity system was based on the distribution of direct current and the distribution of electricity took place only over short distances (hundreds of metres) using large and massive copper wiring, at low voltage.

The most important reverse salient experienced by this system was the fact that it was only economical to use the electricity in built-up and localised areas due to high transmission losses. This reverse salient could, in principle, be translated into various critical problems including:

- Reduction of transmission losses
- Improvements to the efficiency of smaller power plants (this would allow various independent grids)
- Battery-based distribution in sparsely populated areas

The critical problem that Edison chose was the reduction of transmission losses by an improved grid. Despite an accurate definition of the problem, the inventors and engineers using direct current (DC) at the end of the 19th century were not able to find a solution to this problem.

In the end Edison's competitor Westinghouse found a solution outside the DC system, i.e. alternating current (AC), which enabled the system to transform the voltage readily using transformers. This led to the co-existence of two conflicting electricity systems for some time and competition between them was severe. Edison even invented the electric chair to show the American public that the high voltages of the Westinghouse system were dangerous.[5] But his efforts were in vain and AC took over.

This example illustrates how new systems can form when old systems are unable to resolve their reverse salients through their own means.

- In the **first phase**, the emphasis is on the invention and development of a system. The professionals dominating this phase are inventors/entrepreneurs. They differ from regular inventors in their attempt to organise the entire process from invention to 'ready for use'. Edison, of course, is the supreme example of such a person. Engineers, managers and banks are also important during this phase, but they are of less importance than the inventor/entrepreneur

---

5   M. Essig, *Edison and the Electric Chair: A Story of Light and Death* (New York: Walker & Company, 2003).

- In the **second phase**, the most important process is technology transfer from region to region or from continent to continent. The transfer of Edison's electrical system from New York to cities such as Berlin and London is an example of this. During this phase, various groups become involved in the development such as traditional entrepreneurs and banks

- The essential characteristic of the **third phase** of the model is fast system growth. Growth occurs geographically (e.g. electricity systems spread) as well as quantitatively (e.g. electricity consumption and production rise) and qualitatively (e.g. electricity penetrates many technologies). Reverse salients occur in this growth and are redefined into critical problems

- The **fourth phase** of systems development is characterised by substantial momentum. A system with a substantial momentum possesses mass, velocity and a direction of motion. 'Mass' points to machines, equipment, structures and other physical artefacts into which capital has been invested. The 'mass' also stems from the involvement of people who possess professional skills specifically suited to the system. Entrepreneurial businesses, government services, trade unions, educational institutions and other organisations that are directly attuned to the system's core also contribute to the momentum of the system. Taken together, these organisations form the culture of the system. A system also has a measurable speed of growth. That speed often increases in this phase. A system also possesses a direction of motion, i.e. goals

- The **last phase** of system development can be characterised by the qualitative change in the nature of the occurring reverse salients and by the advent of financers and consultants as problem-solvers. Managers play the leading part in the momentum-gaining phase. In this phase, which is concerned with developing regional systems, the most important reverse salients stem from the need to finance large-scale systems and to clear legal and political barriers

Economic approaches to the creation of new technology

This section presents an overview of the way in which economists think about technology, beginning with the neoclassical economic framework. This framework is mainly useful for analysing substitution of 'off-the-shelf' technologies. Factors limiting substitution such as 'lock-in' are then considered in combination with the quasi-evolutionary approach of technological change.

The neoclassical economic framework

Neoclassical economic theory was outlined in Chapter 4. It still has great influence today and plays a major role in government policies. This section examines only those elements that are the necessary to explain technological innovation.

Neoclassical theory assumes entrepreneurs wishing to produce a certain quantity of a product have a choice between different combinations of production factors. The term 'production factors' when used in economics signifies all that is needed for production:

- Labour

- Monetary resources

- Physical resources

Production can be performed in a capital-intensive way (i.e. using lots of machines) or in a labour-intensive way. Neoclassical theory assumes that the choice depends primarily on the cost of capital (interest) and labour (wages).

This choice is depicted in the production function (Figure 9.7). The function yields all combinations of labour (x-axis) and capital (y-axis) with which a certain quantity (Q) of a product can be produced.

Entrepreneurs are free to choose any point on this curve; where on the curve they decide to be results from the pricing of labour and capital. When certain prices for labour and capital are given, each of the straight lines in Figure 9.7 shows the amounts of capital and labour that can be bought for a certain, fixed price. For curves more towards the right (or towards the top), this price is higher. An entrepreneur who wishes to produce an amount of products Q at the lowest possible cost will choose the point on the curve that lies at a tangent to the cost function. This is the point on the curve where the total costs of labour and capital combined are lowest.

When the pricing of labour and capital changes, the slope of the lines changes and hence displaces the tangent point. This encourages the entrepreneur to switch to a different point on the curve, i.e. to a different distribution of labour and capital. If, for example, labour becomes more expensive, the lines will have a steeper slope and the entrepreneur will move to a point 'higher' up the curve to a more capital-intensive production method. This does not happen instantly, but at a time that suits the entrepreneur, e.g. when machines need to be replaced.

For our purposes, neoclassical theory offers some possibilities but also some important restrictions. The possibilities lie in the fact that the influence of capital and labour costs on technological changes can be determined. Every choice made between capital and labour entails a technological choice, i.e. a choice between production technologies. Using this theory, we can gain some understanding of the influence of labour costs and interest rates on the choice of production technologies. We could even make a more com-

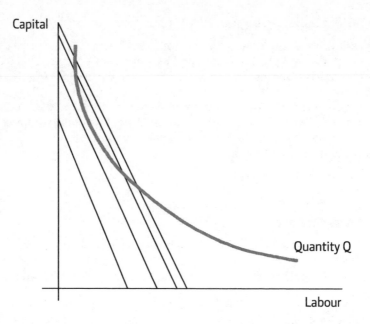

FIGURE 9.7 **Production function**

plex production function if we included a third axis that presented resource costs.

A government sometimes decides to influence factor costs in order to influence the outcome of technological choices. For instance, the government can impose taxes on energy consumption to stimulate businesses to use more energy-efficient production methods. Taxes on emissions can also be used to achieve technological changes. These are examples in which the costs of a production factor are raised artificially to stimulate the adoption of a specific technological solution.

Neoclassical theory discerns between shifts along the production curve and change in position of (part of) the production curve itself. In the first case, there is no innovation—existing technologies are merely being exchanged. In the second case, innovation takes place and the production curve shifts towards the origin of the co-ordinate system. However, this theory does not explain where these innovations come from. It is assumed they come from outside the economic system or in other words are 'manna from heaven'. This is a weakness—economic factors certainly do influence the direction of innovations.

For example, higher electricity prices will not only stimulate entrepreneurs to search for equipment with a low electricity consumption, but will also provide a stimulus for research into more energy-efficient technologies (Figure 9.8).

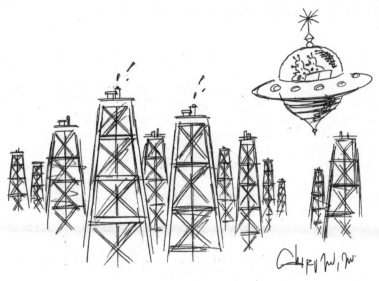

'They're still using that greasy kid stuff'

FIGURE 9.8

Source: www.CartoonStock.com, 11 November 2005; © Cartoon Stock Ltd

A second problem is the assumption that a large choice of alternative production methods is available. In practice, this is rarely the case. Usually, there are only a few alternatives for the small part of the production function around which the prices of capital and labour fluctuate. If those prices start to deviate strongly, all sorts of new technologies will have to be developed. Thus, technological innovations may be needed even for movement along the production function.

'Push–pull' debate

Economists have long argued about what is the driving force behind the occurrence of technological innovations.

The proponents of the 'technology push' theory view developments within science and technology as the main driving force for technological innovations. They share their views with technological determinism (see Chapter 5).

The 'demand pull' theorists mainly see market demand as the main cause of technological innovations. Market demand largely determines the formation and introduction of new technological possibilities.

Representatives from both camps have supported their theories with a host of empirical research. However, these schools of thought have closed in on each other and a number of economists now see the combination of technology push and market pull as the driving force behind technological innovations.

The evolutionary theory of technological change

The production function is not a particularly satisfactory method for analysing the economics of innovation and, in the 1970s, economists moved in a new direction. The development of new technology always depends on the monetary state of technological knowledge and skills, and not just on economic parameters. New technologies emerge as small variations and the selection of these variations occurs in society. This is an **evolutionary process.**

These variation and selection processes do not occur at random or for no reason, but show a clear structure. A certain rigidity and inertia is present in the rate of change of technology, which prevents variations from popping up without limits. The phrase '**technological regime**'[6] is used for this.

A technological regime forms the framework of reasoning for technology developers. It consists of a limited number of:

- Scientific principles
- Practical approaches
- Rules for seeking solutions to problems (heuristics)

Central to the technological regime is the exemplar. The exemplar is the basic example—the basic technological structure that all innovators try to improve on. For example, the DC3 was the exemplar for civil aircraft development from 1940 to 1960.

The direction to search for improvement is determined by heuristics, i.e. rules for searching such as aiming for weight reduction in aircraft development. Technological regimes are often stable over long periods. For example, the jet aeroplane, the mainstay of intercontinental air transport, had come into existence by the end of the 1950s. The turbojet propulsion, a closed aluminium fuselage and its shape have hardly changed for over half a century. Variations are sought in materials, electronics and turbine efficiency, and small adaptations in aerodynamic design.

If a technology is guided by a regime for several years, the subsequent forms of the technology create a technological trajectory, i.e. a path that describes the 'direction of progress' within that technological regime.

A technological regime and the direction of development of technologies are, to a certain extent, inert and subject to little change. It does not mean

6 Sometimes also called 'technological paradigm'.

that the current regime is abandoned even when it becomes apparent that the problems receiving attention can be dealt with only in a less satisfactory way than they could be in a different regime. A regime is seldom abandoned, mainly because new regimes often do not create very efficient technologies in their early stages. Radical new technologies are often only introduced when everybody is convinced of their future success.

Technological regimes are crucial for technology development. On one hand, they restrict the development of technology by imposing limitations on the possible variation in new technologies. On the other hand, regimes enable accelerated development by concentrating effort and available resources in one specific direction of research and development.

Not every new technology is successful. The term '**selection environment**' is used to indicate the collection of stakeholders, structures and institutions that determine the selection process. Selection is not the same as market selection. Existing technologies in the market, legislation, insurances and available infrastructure may all play a role in the selection of technologies.

Quasi-evolutionary theory

The relationship between technology development, variation and the selection environment is a complex one. The influence of the selection environment on the development process occurs not only after technologies have been developed, but also during its development. An example of this is the anticipation of the increasing demand for ecological products by the research department of a food company.

Variation is not the only thing influenced by the selection environment. The reverse also occurs. The variation process changes the selection environment, e.g. in cases where new technologies create completely new options. This happened, for example, with the introduction of mobile phones, which enabled people to adopt new lifestyles—a change to which advertising contributes.

Because of this coupling between variation and selection, the term 'quasi-evolutionary theory' is often used nowadays. The selection environment is often anticipated in the variation process, and the selection environment is often strongly influenced by the variation process.

Non-linear effects in the diffusion of technology

Once technologies have been developed, they are introduced into society. If successful, they gradually diffuse into society. However, this diffusion process is often rather unpredictable. The mobile phone conquered us all in a few years, while other technologies (e.g. electric cars) appear to be in the doldrums forever. This is not just fate. If we look more closely, some interesting mechanisms can be discerned.

'Lock-in'

Windows and the car with an internal combustion engine are products for which it is hard to imagine alternatives. It seems as if there have never been any alternatives to them, as if these products possess a superior quality over any other possibilities. Still, they often are no more than the 'lucky' winners of a competitive battle.

Take MS-DOS, for instance. If the wife of Gary Kildall (the inventor of cp/m)[7] had not sent away the IBM managers who rang her doorbell sometime in July 1980, they probably would not have driven to the Microsoft offices and this book would perhaps have been written using a version of cp/m.[8]

Economists call this the 'winner takes all' principle. With technological standards, there is one system dominating the market. Sometimes, there is room for a second one, but producers of a third or fourth system have very little chance of ever making a profit. A company that has exclusive possession of a standard has a near-monopoly. In this way:

- Microsoft holds 90% of the market in PC operating systems

- Intel holds 90% of the market in microprocessors

- IBM holds 83% of the market in mainframe computers and practically the entire non-Japanese market for mainframe operating systems.

A classic economic law predicts that returns decrease as sales increase. According to that law, a product whose sales volume keeps increasing reaches a ceiling beyond which profit does not increase any further. The causes are increasing costs per unit sold and the launching of alternative products by competitors. This mechanism keeps prices low and prevents excessive market domination.

However, this law does not hold in the case of modern technologies where the situation is one of increasing returns. Sales start off slowly but, when in the end everyone switches to the dominant technology, market shares soar towards 90%. After this phase, any potential competitor faces a considerable challenge.[9] This is reinforced by companies that start using the standard to supply additional services. For example, computer game developers want to develop games only for the main platforms such as PlayStation®, Xbox and

7 An operating system for 8-bit computers. See, for example, www.gaby.de/ecpm. htm, 11 November 2005.
8 Examples from: G. Bakker, *Intermediair* 33.21 (22 May 1997): 47-51.
9 W.B. Arthur, 'Increasing Returns and the New World of Business', *Harvard Business Review* 74.4 (July/August 1996): 100-109.

GameCube. They thus reinforce the standard as these platforms become more popular.

Users desire a product of a uniformly valid standard so they can obtain compatible software and exchange information with others easily. This high 'networking' tendency means that a small advantage in the early developments of such a standard will persuade more and more people to buy the product, after which sales will increase by themselves. Production costs drop, while those of the competitor rise. We are **'locked in'** as the standard is very hard to change. The change from leaded to unleaded petrol, which took some 25 years, is one of the few examples of change from a locked-in technology. The duration of this substitution indicates the difficulties involved.

Technological standards often enter the market in pairs—a software part and a hardware part. The software part anchors itself in the market and gives the hardware part leverage to increase sales. When consumers have bought the software part, they are committed to that standard. For example, no-one will decide to exchange all the Windows machines in an office for Apple computers if that implies that all software must also be replaced.

The heavy competition amplifies the snowball effect. Manufacturers lower their prices and thus more people buy their new products, creating an even larger market for the winner. For example, Netscape and Microsoft even gave away products to win the battle for the Internet browser market, with the result that it expanded rapidly in the 1990s.

Killer applications

There can only be one winner. But who is it going to be? A business interested in making its technology the standard on the market should not seek to optimise short-term profits. For example, Philips gave away licences in the 1960s to everyone wishing to produce its compact music cassettes. In the 1980s, Microsoft signed an agreement with IBM that did not yield much profit, but ensured that MS-DOS became the standard operating system for PCs.

A new generic technology needs a 'killer application' to gain market acceptance. The killer application encounters such popularity that it persuades millions of people to buy the product. For example, the spreadsheet program VisiCalc was largely responsible for the successful introduction of the PC by Apple and IBM.

Edison invented the phonograph using wax rolls in 1877. Edison, who was more of an inventor than an entrepreneur, thought that the novelty of his product lay in its many uses—as a way of a voice recorder for recording news, speeches, strange languages and also maybe music. The German entrepreneur Emile Berliner foresaw music becoming the killer application. He marketed the gramophone he had invented and, at the same time, started a record label. Within a few years, the gramophone was the market leader and Edison started producing it too.

QWERTY: the most boring standard ever

The QWERTY standard is the classical example of a standard that is impossible to replace. The typewriter invented by the American Samuel Sholes suffered from two problems:

- The little hammers got stuck together when he tried to type fast
- The machine needed to have a special trick he could demonstrate

By using the QWERTY configuration, Sholes was able to make the most often-used keys widely spaced so that the hammers would not get in each other's way. It also made it possible to type the word 'typewriter' very quickly using only the keys on the top row—a handy sales trick.

All Sholes then needed was someone to produce the machines. He found an ally in arms producer Remmington. To help sales of the typewriter, Remmington organised typing contests, which competing machines could also enter, and then gave contracts to the winners.

It was only a short while before typists invented touch-typing using ten fingers. Educational institutes quickly adopted the system and soon everybody wanted the QWERTY system:

- Companies because their secretaries could work quickly using the system
- Educational institutes because most companies used QWERTY machines

Even though that there was no longer a technical necessity for the QWERTY arrangement, most other manufacturers had also adopted the system by the beginning of the 20th century.

The most well-known alternative to QWERTY is the keyboard developed by August Dvorak in the 1930s. Using the 'Dvorak Simplified Keyboard', people learned to type twice as fast as QWERTY people and experienced a 20-fold decrease in hand strain. Some Dvorak-machines were made and the configuration is optional on some Apple computers, but Dvorak was not a success. QWERTY had installed itself into the brains of millions of people and could not be erased from them.

Why has QWERTY not been replaced? Surely an ergonomical layout would have prevented lots of stress and maladies such as repetitive strain injury (RSI)? And it can't be that expensive to plug a new keyboard into your computer?

A number of explanations can be given for the survival of the QWERTY system. First, touch-typing using 'QWERTY' is a skill that is impossible to forget, in the same way as swimming and riding a bicycle. People who touch-type will therefore not be quick to switch to a different keyboard layout.

A second explanation is that the improvements in learning speed, typing speed and hand strain are not apparently sufficient reason for young people and typing institutes to justify switching to a new keyboard layout. Typing is supposed to be difficult and hard to learn. Experts on innovation state that a new standard should be about ten times as good as an old one if it is to be adopted.

The laserdisc is a more recent example. Philips has tried, with no avail, to market such a device for playing interactive CDs under different names three separate times. Competitor Pioneer introduced the killer application—a laserdisc machine suitable for karaoke, which is incredibly popular in Japan. Japanese sales skyrocketed, as did Pioneer's market share.

A lack of killer applications can be worrying. For example, ISDN digital telephony only took off when the growth in interest in browsing the internet made people request multiple telephone lines. Companies selling home systems (i.e. systems that integrate various electrical appliances in the home) are still searching for an application that will convince people to buy their systems.

The mechanism behind lock-in: positive feedback[10]

Over time, technological trajectories or 'paths' form. The technologies in which companies or economies get stuck do not always have to be those that are most efficient for their users.[11] A market advantage at the right moment may lead to an advantage for one of the competing technologies that reinforces itself; as the product becomes more attractive, the more it is sold.

The formation of technological paths, which can be inefficient, is a consequence of the phenomenon 'increasing returns with increasing market penetration'. This means that the more a technology is adopted, the more it improves and the more attractive the technology becomes for further development. This situation is called a '**positive feedback**'. Six factors lead to positive feedback:

- **Expectations.** The development of a certain technology can be influenced and accelerated by the expectations people hold as to its success. Nobody wants to buy a product that they think is going to be withdrawn from the market

- **Familiarity.** When a technology is better known and better understood, it has an increased chance of being adopted. This factor has also been described as 'increasing returns by information'

- **Network characteristics.** It is advantageous for a technology to be associated with a network of users because this increases availability and the number of product varieties. A good example here is the VHS video-recorder system. To be able to function, this technology

---

10 Revised extract from: C. van Leenders and P. de Jong, 'Milieustrategieën en positieve feedback: kunststofverpakkingsafval als illustratie' (Amsterdam: University of Amsterdam [UvA], 1996).
11 In a neoclassical economy, it is assumed that the 'best' technology will conquer the market. In Arthur's model, the process is trajectory-dependent, which makes it impossible to predict which technology will conquer the market. One consequence is that an inefficient technology can actually win the race.

Video-recorders

Before 1974, only professionals could record TV programmes and the equipment was voluminous and expensive. Large electronic companies were all working to develop video-recorders for the consumer market. In 1974, Sony was the first to introduce a consumer video-recorder, using its Betamax format. Sony offered its competitor JVC a licence, but JVC refused to pay and introduced its own system, VHS, in 1976. In 1975, Sony had a monopoly and sold 30,000 Betamax video-recorders in the USA for about $2,000 each. Philips introduced its own Video2000 (V2000) system. These systems were not compatible, i.e. the recording tapes were specific for each system. Market insiders expected the three standards would all conquer part of the market such as had happened with records (33, 45 and 78 rpm).

Sony's Betamax was first, but JVC's VHS video-recorder had tapes that could record almost two hours of programmes instead of the one hour offered by Sony and Philips. In 1977, the film companies Disney and Universal saw the video-recorder as a threat and attempted to obtain copyright licence fees from the sales of blank video-tapes. This attempt failed and so they decided to sell pre-recorded video-tapes of their films themselves.

In 1978, the VHS system was further improved to handle tapes of almost three hours' recording length. Within a year Betamax and V2000 could do the same but is was too late. A new phenomenon had arrived—the video-rental store. They popped up everywhere in the USA and were filled with VHS movies. Many were adult movies that appeared to be more popular viewed at home than in the cinema.

The US market share of JVC video-recorders peaked as Sony's plunged. Philips tried to improve on its negligible market share by lowering prices. JVC followed, but was able to reduce its costs more than the others due to economies of scale. Gradually it became clear that Sony and Philips could not catch up. Consumers preferred a video-recorder for which they could obtain the most video-tapes (from friends or from rental stores). The video-rental stores also had limits on how many tapes they could hold, so they offered those that were most in demand and these were VHS types. After years of desperately fighting, Philips gave up in 1985 and Sony in January 1988.

But were Betamax and V2000 inferior products? Many video enthusiasts still want V2000 or Betamax and they exchange tapes and spare parts on the Internet. So, despite the fact that nobody spends any money on the technical innovation of these systems, there are people that believe in their superiority.

VHS was the winner not through its technical merits or its price. Consumers picked the video-recorder that their friends had so they could exchange tapes and would have biggest range of choice when renting a video. The more people bought VHS, the more attractive it became for the customer, i.e. 'positive feedback'.

needs a network consisting of video-rental stores stocked with VHS tapes. The more users there are, the more possibilities there are for users to obtain full benefit from their type of video-recorder

- **Technological connectivity.** Feedback processes are stimulated by the occurrence of 'technological connectivity' or complementary technologies: a technology that fits into the system of already exist-ing technologies has a relatively better chance of developing than a technology that lacks those connections

- **Economies of scale.** Increasing sales of products lowers the costs per unit production. This means that a technology can become more economical when applied on a larger scale

- **Learning.** Positive feedback during the development of a technol-ogy can be induced by learning. A technology can be improved more quickly when more is learned during its use. When more is learned about a technology, this technology gains an advantage in applica-tion. When a company learns more about using a specific technol-ogy while learning little about another, the latter has less chance of being adopted

Technology policy

Chapter 4 discusses policy options for working on sustainable development, while Chapter 5 discusses the role of technology in sustainability problems and the question of whether the course of technology could be influenced. The answer is positive, though there are periods in which specific technolo-gies can rarely be influenced. But can we use these concepts to steer tech-nologies towards sustainable development?

Social constructivism

The social constructivist model emphasises the influence of various actors in the development of technologies. If choices regarding new technologies are made by social processes, we might be able to 'democratise' technological decision-making or, at least, open the decision-making process to 'green' stakeholders.

Sometimes this requires policy measures to give specific actors more access to the decision-making process. Reinforcing the position of environ-mental non-governmental organisations (NGOs) in technological decision-making can influence the technological outcome, e.g. patient organisations can play a role in decision-making on medical technologies.

Stimulating specific relations, such as between universities and small and medium-sized enterprises (SMEs) can influence research in universities and stimulate innovation in these companies. Science shops are university entry points that help NGOs access scientific expertise and can help them engage in the expert debates regarding new technology.

Systems approach

In the systems approach, technological change occurs as a reverse salient becomes evident in the system and is translated into a critical problem. If no clear critical problem exists, reverse salients can remain 'slumbering' for a long time, e.g. congestion has been a slumbering reverse salient of the transport system for a long time.

The availability of a clearly defined critical problem enables the divesture of larger resources towards the problem. Policies could be directed towards:

- Stimulating basic research to develop options for critical problems
- Stimulating and facilitating the formation of consensus regarding critical problems

Policies could target:

- The external conditions for systems growth/decline
- The reverse salients and the choices that are made in their redefinition as critical problems

Quasi-evolutionary economics

In the quasi-evolutionary theory, technological change could be influenced by variation and by the selection process.

- **Variation.** Policies may aim to generate a greater variety of technological options by stimulating or creating facilities for fundamental scientific research or more specifically for technological mavericks, i.e. scientists not in the mainstream

- **Selection**. New technologies have to compete with established technologies, which have been optimised for a long period. They are frequently therefore unable to compete in the market as they often contain various practical drawbacks. A period of protection, during which the technology may be improved by learning in practice, might help. A specific part of the selection environment—a **'niche'**—may provide such a protected environment. For example, buses using cleaner fuels may be used first by metropolitan transport companies because the company itself can refuel the buses and the buses will remain in a limited area. There are also direct benefits (cleaner air) for an important stakeholder group—the area's residents, who will probably have to pay higher costs. Creating and managing of niches for new technologies is therefore an interesting instrument

The selection of technologies can also be influenced more directly. Cleaner technologies often have to overcome the problem that their production in low numbers makes them expensive, e.g. photovoltaic (PV) cells. Only a few people buy them and so it is not viable to build the larger-scale production units that would enable cheaper production. A temporary subsidy or government purchase of the first products can help to overcome this hurdle.

Variation and selection are not completely independent. People working on the creation of new technologies already carry out part of the selection and it is important that they receive the right signals. Therefore, it is essential to stimulate institutions (so-called **nexus**) that could convey the signals from the selection environment into the variation environment. For example, insurance companies (covering liability claims for environmental hazards) might be important in feeding back possible future liabilities into the variation process of companies.

Variation processes take time. Governments could set targets for future selection of technologies to give companies the opportunity to work towards these targets. This is called **'technology forcing'**. This appears a clear-cut measure, but it often creates much hostility between the legislature and the industrial sector involved. For example, the US federal government tried to force motor manufacturers to produce cleaner cars with the Clean Air Act 1970. However, the result was a harsh legal battle and the reduction of car emissions was no more than could have been expected without these measures.[12]

---

12 K. Mulder, 'The Effectiveness of Technology Forcing, CFCs and Cleaner Transport', in Proceedings of the International Conference on Technology Policy and Innovation held in Lodz, 6–8 July 2005.

Lock-in and positive feedback

In the history of specific technologies, minor factors can be very influential at the right moments while it can be almost impossible to change anything at other junctures. It is therefore important to focus policies at the point when new technologies are locked in. However, unlocking is generally only possible at high cost.

Technology transfer

The creation of technologies is dependent on various local conditions. For example, you sometimes cannot switch electrical equipment between countries as their grids deliver different voltages or frequencies. But, when transferring technologies between quite different cultures, differences in local culture, organisation and availability of infrastructure may be far more important.

This effect is evident from the theories outlined above, which all emphasise the effects of the social environment on the creation of technologies through:

- Relevant social groups

- The system's environment

- The selection environment

Thus, technology transfer between societies presupposes that the societies are more or less equivalent. Adoption of a technology is not a neutral choice but implies also a choice to change your—and others'—way of life. Many people do not recognise this point. For example, the adoption of technology is a key issue in the Amish communities in the USA:

> The Amish, Anabaptists or re-baptisers, originate from the French Alsace. They hold a literal interpretation of the bible, which denies them the use of any violence. They left France after Napoleon tried to force them into his army. Their communities are now concentrated in Pennsylvania. The social life in Amish communities is guided by strict rules that aim at separating them from worldly life. Various technologies such as cars or televisions are ruled out. Some modern technologies are accepted at their semi-annual meetings. The principle behind decisions is if the technology will affect the Amish way of life. Naturally, the external world has some requirements, and therefore the Amish allowed electric milk-stirrers, and welding, but only if the electricity was produced by a local genera-

tor. To prevent the use of 'worldly' equipment, all other electrical devices can only be used with 12 volts.[13]

Various things can go wrong if a developing country imports a technology that is developed in a Western industrialised society:

- The technology is irreconcilable with the values and culture of various social groups. For example, societies of hunters and gatherers (see Chapter 3) tend to have strict rules that prohibit the outright exploitation of nature. Often, they regard (parts of) nature as divine[14] and therefore reject Western agricultural or mining methods

- Technologies are irreconcilable with the order of society. For example, technologies that facilitate distribution of knowledge are often irreconcilable with a strong hierarchy. Medical technologies may pose a threat to the traditional doctor, who often holds a powerful position

- The technology makes sense only as part of a larger system, which does not exist. Equipment needs fuel, maintenance, repair, skilled personnel, etc. Introducing equipment only makes sense if the whole system is functioning well

- Technologies have gradually evolved according to the selection standards of the industrialised nations. For example, the Western customer who is used to high-quality roads and demands certain levels of luxury has dominated the evolution of cars. The Citroën 2CV could be seen as the exemplar of an alternative trajectory suitable for countries with a lower income and poor infrastructure

Sometimes, the transfer of technologies creates a 'Westernised' community within a developing country. In this way, imported technologies can introduce tensions within a society. This leads to the so-called dual society seen in so many developing countries with small islands of urban industrialism surrounded by slums and large areas of poor countryside.

Monuments such as rusting tractors in rural Africa attest to the fact that the transfer of technology cannot work without:

13 J. Sha, 'The Amish: Technology Practice and Technological Change' (1999); www. shawcreekgeneralstore.com/amish_article1.htm, 11 November 2005.
14 The famous 'speech of Chief Seattle' (which might actually never have occurred) illustrates nicely the confrontation of American and Native American value systems with respect to nature. J.L. Clark, 'Thus Spoke Chief Seattle: The Story of An Undocumented Speech', *Prologue* 18.1 (Spring 1985); www.archives.gov/ publications/prologue/1985/spring/chief-seattle.html, 11 November 2005.

- The physical and human resources to sustain it
- Regard for the social, economic and environmental context into which it is parachuted

The concept of '**appropriate technology**' was introduced in the 1970s in response to the failure of technology transfer. Appropriate technology should be adapted to the circumstances of developing nations, i.e. it should:

- Be simple to apply
- Not be capital-intensive
- Not be energy-intensive (i.e. requiring little non-renewable energy to operate, build or maintain)
- Use local resources and labour
- Nurture the environment and human health[15]

Introducing such technologies in developing countries is not easy. Take the example of a manually operated water pump (Figure 9.9). The pump is cheap and easy to operate by uneducated people. It is designed so that it can be repaired and maintained using local materials. If the pump is introduced in such a way that all local farmers can use it to irrigate their land, the technology is appropriate. But, if the pump becomes the property of one large local landowner, the difference between rich and poor is enlarged locally.

Who is to decide? If a developing country expresses the need for a diesel engine pump, should Western organisations decide whether it needs this type of water pump? Should Western organisations decide that the local hierarchy should be ended and local democracy introduced?

The development process should always be steered by the people for whom the development has to be beneficial. It is their society, their culture and their future that is at stake. Foreign organisations should not steer these choices.

The preferred way of international co-operation for development should therefore be to co-operate in technology capacity building such that developing countries can assess, adopt, manage and apply environmentally sound technologies themselves.[16]

However, aid organisations may face some difficult dilemmas regarding respect for local cultures, personal ethics and universal human rights. Should neglect for the rights of women be tolerated? What about racial prejudice? What about locally accepted corruption and nepotism?

---

15 National Centre for Appropriate Technology; www.ncat.org/about_history.html, 11 November 2005.
16 Cf. Chapter 34 of Agenda 21, 'Transfer of Environmentally Sound Technology, Co-operation and Capacity-building'; www.gdrc.org/techtran/a21_34.html, 11 November 2005.

FIGURE 9.9 Manually
operated water pump

© Crelis Rammelt, TU Delft

The whole concept of 'giving aid' is rather patronising. We must seek dialogue. The cultures that are often called primitive may be able to teach the world valuable lessons in how to live happily without ruining the life-supporting systems of our planet. The industrialised world may claim to have developed superior technologies, but it does not have the credentials to claim to be more knowledgeable in regard to developing in a sustainable way.

Questions, discussion and exercises

1. What is a reverse salient and what is a critical problem? Define them for the Panama Canal system (see over).

    Could you answer the same question for the world shipping system?

2. Necar was the first vehicle powered by a fuel cell. It is a Mercedes-Benz A-class with the same space for passengers and luggage as a standard model. The fuel cell provides the electricity for the electric engine that propels the car. Hydrogen for the fuel cell is produced from methanol carried on board; the necessary space for the hydrogen tank and other equipment could be obtained by equipping the car with a sandwich floor. Necar can also refuel at gas stations. The fuel-cell car has been extensively tested in normal everyday use since 2003.

    Describe how 'positive feedback' could occur when this vehicle is made available and how it could influence the acceptance of the fuel-cell vehicle.

Panama Canal system

The Panama Canal is too small for modern post-Panamax vessels. In 2002, the Panama Canal Authority asked two organisations to design new larger locks next to the existing locks. The US Army Corps of Engineers was asked to design locks for the Atlantic entrance to the Canal near Christobal and a consortium led by the Belgian company Tractebel Development Engineering was asked to design locks for the Pacific entrance at Balboa.

Provided the canal itself is deepened from 13 to 17 metres, the new locks will allow much larger ships to use it. If they are built, the new locks will be the largest in the world and will cost US$6–8 billion.

The main problem is the availability of water to replenish that lost from the canal each time the locks open. Lake Gatun, the central reservoir for the canal, is already being enlarged and studies have been commissioned for a new reservoir. Flooding of part of the Panama jungle has been rejected. Studies are also to be undertaken to reduce the water loss from the Canal during each lock opening; US company WPSI has produced a plan to pump the water from the locks back into Lake Gatun.

3. Discuss the pros and cons of the exports of agricultural machinery to a specific developing country. Study the state of agriculture, its culture and its economic organisation. What would you do if your judgement regarding this issue was challenged by a request for equipment from the country that you have chosen?

10 Technology for sustainable development

In Chapter 1, we explained why the world needs to develop in a sustainable way and why we need to improve the environmental efficiency of our production and consumption by a factor of 32.4. This number indicates that we have to think in leaps and not in marginal improvements. In Chapter 2, we discussed the problems of the natural systems that sustain human life and, in Chapter 3, the issue of development. The roles of governments and companies in the problems of sustainable development were discussed in Chapters 4 and 7, respectively, while Chapter 6 considered how to measure sustainability. Chapter 8 discussed the technological design process and Chapter 9 technological innovations. Chapter 5 discussed the role of technology in the history of industrialised society.

In this chapter, we present some insights, tools and principles that may lead to leaps in the environmental efficiencies of technology. We deal first with environmental technologies and then discuss the characteristics of sustainable innovations.

In Chapter 9, we discussed innovation mainly at the level of artefacts. For sustainable development, innovations at systems levels are also needed. Systems innovations do not only change specific products or processes, but the configuration of the system of which these products or processes are part.

But systems innovations are not enough. Technological leaps will be needed that will also demand adaptation of consumer behaviour and restructuring of the organisation of society. These types of innovations are called **transitions**. They do not occur very often and, in general, we recognise them only afterwards. For example, the invention of the transistor and the integrated circuit together with the development of more complex forms of social organisation lead to the advances in information and communication technologies that have revolutionised our society in the past decades.

Environmental technologies

Environmental problems come in all shapes and sizes. In finding solutions to these problems, it is evident that people have aimed first at the easier solutions—the low-hanging fruits.

Ever since the Neolithic agricultural revolution, humankind has been confronted by environmental problems.

- Historically, the first environmental problems were caused by the dense population of the first cities. The easiest way to tackle these problems was by using triple D technologies (see Chapter 7)

- As pollution became a problem, people began to think about pollution prevention (see Chapter 7)

- Residual waste can be treated using end-of-pipe technologies. Such technologies prevent/reduce emissions but leave the production system unaffected (see Chapter 7)

- Restoration technology is a specific kind of end-of-pipe technology. We have an obligation, at the very least, to clear up the worst pollution from the past and to isolate polluted sites from their unpolluted environment

- In many cases, however, the preferred option is to reduce the environmental burden by changing the production process. In this way, further reductions in pollution and resource consumption can be achieved. Complete redesign of production processes can lead to both environmental gains and cost reduction

- Ultimately, we must develop technologies for sustainable production and consumption because none of the technologies above will suffice to solve the environmental problems we face

Sustainable technologies go beyond environmental technologies. While environmental technologies are concerned with producing goods and services with minimal pollution, sustainable technologies have a far broader aim. This is to enable us to fulfil the needs of the whole humanity without:

- Exceeding its ecological recovery capacity
- Consolidating or promoting inequity

Characteristics of sustainable technologies

A sustainable technology means more than merely producing goods without pollution or ecological destruction. Sustainable technology means fulfilling people's needs in such a way that the recovery capacity of the planet and the recovery capacity of local ecosystems are not exceeded. The aim is to bring the world's use of natural resources within the boundaries set by the Earth's recovery capacity. What does it take to achieve this?

Fulfilment of needs

The first step in developing technologies for sustainable development should be to analyse the need that is fulfilled by the product. However, it is important to recognise that consumers may have 'hidden' needs that are not necessarily articulated. Companies or governments may also have hidden needs such as national prestige or the status of officials.

Specific products often raise an issue of legitimacy (e.g. should we allow Formula 1 races, cigarettes, etc.). However, these products reflect a need that is, as such, legitimate.[1] The challenge is to develop more sustainable alternatives to fulfil these needs.

Can a need be fulfilled in a different way? If so, is this alternative preferable in environmental, social and ethical terms?

Thinking in fulfilment of needs often requires a crossing of disciplinary boundaries. The best solution for a problem may be outside the discipline in which you are trained. Disciplinary training can sometimes prevent you from making a leap and limits the scope of alternatives that we consider.

For example, melting a metal requires a minimum amount of energy, no matter how efficiently a process is designed. If further improvements are sought, efficient recycling schemes contribute to reduce the demand for ores and energy consumption. For some applications of a metal such as zinc, recycling is impossible as much of the metal is dissipated during use. In the case of galvanised steel, for example, the zinc layer dissipates into the environment. Further progress by means of technological innovation would therefore seem impossible until it is realised that the user does not necessarily want a galvanised steel product! What a user wants is a durable product that renders him or her a specific service.

Brainstorming sessions to generate alternative modes of need fulfilment are very important in opening our minds to alternative ideas. If brainstorming is carried out vigorously, then many alternatives normally appear that originate from entirely different sectors. Alternatives for galvanised metals, for example, might include treated wood, painted metals (non-corroding metals), plastics or even a completely redesigned product that renders all these options superfluous. It may even be the case that the manufacture of the product can be superseded by providing a service, e.g. replacing a vehicle used for transport by a means for communication. Such radical innovations involve a new production system as well as new marketing channels and new forms of consumption. But the precondition for arriving at these

1 In the examples given, the need is for a 'thrill' and for 'comfort', respectively.

alternatives is to focus on fulfilling a need, not on improving an existing technology.

Thinking globally

Fulfilling needs in a far more efficient way does not necessarily lead to sustainable solutions. Technologies that are very environmentally efficient may have:

- Various negative side effects
- Longer-term effects
- Limited or small-scale applicability

A number of examples are given below.

- Monsanto's Roundup technology[2] enables farmers to obtain a higher yield with lower use of herbicides and at lower cost. However, the genetically modified corn may affect ecosystems and farmers have to buy the seeds from Monsanto each year. How does this affect the power of farmers *vis-à-vis* agricultural industry?

- Food aid for regions with high malnutrition rates may keep food prices down. This can be important in preventing starvation but, in the long run, will contribute to the degradation of local agriculture

- Highly efficient technologies may be too dependent on a resource that could become scarce in the future. For example, various precious metals and rare earth metals that are used to catalyse the breaking down of toxic materials could become scarce merely through a relatively small extra demand on the market

- Antibiotics are important in healthcare, but their wide-scale use creates resistant bacteria. In the long run, this could create an enormous health risk

Innovation for sustainable development therefore demands a wide view. We have to act locally but evaluate our technologies globally and with a long-

2 Its key elements are a herbicide and a genetically modified corn that is resistant to the herbicide. They are sold in combination.

term view. Moreover, technology assessment is crucial,[3] as are technologies that are aimed at contributing to the common good.

Looking for long-term solutions

Small improvements in environmental efficiency are attractive, but are not enough. We must aim at making leaps. However, a dilemma often occurs. We may improve on unsustainable technologies, but this will not lead us to the ultimate sustainable technology (no depletion of resources, clean, safe, etc.). Should we spend precious research and development (R&D) money on the small improvement options that can be developed quickly or should R&D be dedicated to creating real leaps but which could take considerable time?

In practice, the dilemma might take the following form:

- Should we aim at developing optimised coal-fired power plants or prefer investments aimed at improving wind turbines?

- Should we develop large, efficient and clean municipal solid waste (MSW) incinerators or reduce waste by prevention and recycling?

- Should we develop more durable houses (that are inflexible, need more materials but consume less energy) or should we aim at less durable houses (that are more flexible, need fewer materials but consume more energy)?

The answers are not easy to give. The improvement of existing technologies is often less risky than aiming at breakthrough technology. We could compare the various environmental, social and economic aspects of an investment in an improved coal-fired power plant with the effects of developing an offshore wind park for the average life-span of these technologies. But what about the lasting effects of the creation of new knowledge and new technologies, and the learning that takes place? You always learn from developing a completely new technology (even that the idea does not work in reality), but what is the value of this learning?

The dilemma between short-term improvements versus long-term sustainable technologies can be solved only by assessing all relevant aspects in a long-term perspective. Decisions cannot be made by calculation alone as they always involve a choice for the kind of society we want to create for the future. Recognising the dilemma is crucial in order to be able to make conscious decisions.

Figure 10.1 summarises the challenge for sustainable innovation. We should seek for a wider perspective in time, but also with regard to the stakeholders that we want to take into account. The first environmental improvements generally focus on manufacturing. Methods such as cleaner produc-

---

3 See Chapter 4.

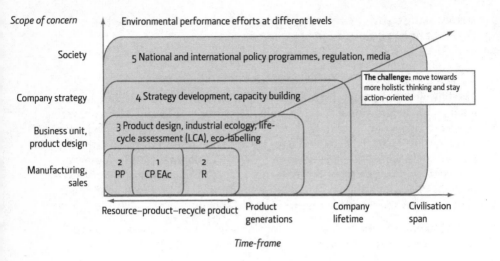

Scope of concern — Environmental performance efforts at different levels

Society — 5 National and international policy programmes, regulation, media

The challenge: move towards more holistic thinking and stay action-oriented

Company strategy — 4 Strategy development, capacity building

Business unit, product design — 3 Product design, industrial ecology, life-cycle assessment (LCA), eco-labelling

Manufacturing, sales —

| 2 PP | 1 CP EAc | 2 R |

Resource–product–recycle product | Product generations | Company lifetime | Civilisation span

Time-frame

CP = cleaner production; EAc = environmental accounting; PP = pollution prevention; R = recycling.

FIGURE 10.1 The challenge for sustainable innovation

tion and environmental accounting (1) should be complemented by pollution prevention and recycling (2). However, these methods should be complemented with environmentally optimised product design and industrial ecology (3). This should be embedded in corporate strategy (4) and national policies and have a long-term perspective (5).

Systems innovations

Systems innovations are innovations that change the structure of a technological system. However, we can discern systems at various levels. For example, the combustion chamber of an engine can be considered a system, but the engine itself or the car that it propels is also a system. We could even analyse the traffic system that encompasses cars, roads, fuel supply, legislation, traffic police, insurances, taxation and maintenance. In this chapter we consider technological systems and systems innovation at a high level— the system involves all those elements that contribute to produce a product or service for a consumer.

An example of systems innovation is the replacement of steam-powered traction for rail transport by diesel- or electrical-powered traction. It encompassed not only replacing locomotives but also all arrangements for energy supply to the trains, maintenance and the training of personnel. Water supply for the locomotives could be abandoned and there will also be major consequences for train timetables.

Systems innovations require large investments and are always destructive for the parts of the system or the whole system that they supersede. For example, the telegraph system was almost destroyed by the telex. Later, both technologies were swept away by the introduction of the fax.

As a consequence of this destructive nature of systems innovations, actors connected to existing production systems are often opposed to the introduction of new systems (Figure 10.2). People may point towards the risks involved and the 'cannibalising' effect of the innovations (i.e. the company loses its existing product). However, as Sony founder Akio Morita once said:

> A company is better off cannibalising its own technologies than having somebody else do it for them.[4]

'Frankly, I'd expected rather more from interactive television'

FIGURE 10.2

Source: www.CartoonStock.com, 14 November 2005; © CartoonStock Ltd

4 E.M. Rheingold, *Made in Japan: Akio Morita and Sony* (New York: NAL Penguin, 1986).

As end-of-pipe technologies have little or no overall impact on the industrial production system as such, they are generally the preferred solution of industry when faced with tackling an environmental crisis. However, adapting the overall production system may often represent a far more effective solution, as it may prevent the creation of waste/pollution instead of dealing with it at the end-of-pipe before it is discharged.

We will examine three examples of systems innovations in more detail:

- Changes in the primary energy base and improving the energy efficiency of the production system

- Changing the raw materials base and re-using discarded end-products

- Avoiding by-products or captive re-use

Changes in the primary energy base

The coming decades are likely to see major changes in the sources of primary energy used in the industrialised nations. Innovations in the energy system itself are dealt with in a later section.

There is considerable potential for reducing energy consumption. Heating and cooling consume significant amounts of energy that can often be reduced by smart heat exchange. For example, data servers need cooling because they are heated by the enormous numbers of microprocessors that they contain. Developing microprocessors that consume less energy can therefore have double benefits.

The noxious emissions from industry and power stations need to be reduced. One way to achieve that aim is to switch from coal- or oil-based fuels to natural gas; this is not sustainable, but still an improvement because less nitrogen oxides (NO_x), sulphur dioxide (SO_2) and particulates are emitted.

Production processes can be adapted to use electricity as a primary energy source. In some regions, a considerable number of vehicles are electrically powered; critics call these 'emission elsewhere vehicles' rather than 'zero emission vehicles'. Still, switching to electricity as a primary energy source may have environmental benefits:

- Power plants burn fuel much more efficiently than other industrial processes. However, part of the energy (3–15%) is lost in transmission and distribution of the electricity via the grid

- A better-developed electricity grid will create more opportunities for producers of renewable energy to sell their energy

- The pollution of one power plant chimney can be far better controlled than that of numerous other exhaust pipes

- Energy users often pollute the air in, for example, urban areas where it is more harmful, while power plants can be sited in an area where they will cause least damage

- A robust and well-integrated power grid with a high-voltage backbone is better able to absorb the irregular supply from various sustainable energy sources. Hydropower dams could then be used as reserves to stabilise the grid and to create a backup capacity

For some purposes, however, conversion to electrical power as a primary energy source is not to be recommended. For example, high-efficiency boilers for domestic heating consume far less primary fuel than electric heaters do indirectly.

The highest efficiencies can be obtained by combined heat and power (CHP) (cogeneration of heat and electricity). In the near future this will probably also be done domestically, thus revolutionising the electricity grid as everybody could be a producer as well as a consumer.[5] But there are numerous technical as well as regulatory problems to overcome before we all become domestic cogenerators of heat and power.[6]

Energy efficiency can be increased enormously by, for example:

- Applying thermal insulation

- Using heat exchangers to pre-heat raw materials and fuels, and simultaneously cool emissions

- Redesigning industrial processes to prevent heating and cooling processes working in opposition

Changing the raw materials base

Changing the primary raw materials base of today's industrial production processes could help reduce waste, pollution and the depletion of non-renewable resources significantly.

Industry produces various wastes that are, in principle, re-usable but which are not re-used for technical/economic reasons (e.g. no cost-effective process available). The European Union (including the new Member States) produces about 1.3 billion tonnes of waste each year, which is about 3.5 tonnes of solid waste per inhabitant.[7]

The recycling processes that are available may become cost-effective as a result of changes to legislation or changes in the market prices of raw materials. Non-availability of cost-effective technologies is thus seldom a pure fact of nature but is also a social fact. The true costs of the waste from a pro-

5 Cf. www.mech.kuleuven.be/tme/research/energy/topics/cogen_en.phtml, 14 November 2005.
6 K. Jackson, 'Domestic Cogeneration', *European Power News*, May/June 2003: 20-21.
7 www.europa.eu.int/comm/environment/waste/index.htm, 14 November 2005.

duction process are often not paid by the company responsible for producing the product. For example, the packaging will become domestic waste once the product has been purchased by a consumer. The producer could decide not to use packaging or to use different packaging. If the costs of waste processing were included in the price of the product, producers would be encouraged to use less packaging material. Waste avoidance (e.g. by changing or reducing packaging, or by changing product design) could reduce domestic waste significantly.

The costs of waste are enormous. For example, every German citizen pays €80 each year for waste processing.

The re-use and recycling of products and/or materials is essential. This can be carried out at various places in the material cycle as a:

- Consumer product
- Component (databases and quality control systems are important for product/part re-use)
- Material (e.g. glass and paper recycling)
- Compound (e.g. waste plastics being converted into feedstock for plastic production)
- Energy (organic materials being burned to recover their energy content)

Even though a substantial proportion could be re-used or recycled as a raw material, consumer products that end up in the dustbin or trash bin are generally landfilled or incinerated. But sorting and cleaning technologies must be further developed before these materials can be efficiently re-used.

A new line of thinking to prevent waste in industrial production is **industrial ecology**. This is based on the notion that industry should follow nature's example, i.e. produce with no net waste. Industrial symbiosis is a micro-level form of industrial ecology (cf. Chapter 7).

With respect to the recycling of domestic waste, efforts so far have focused on metals, glass and paper. Plastics are recycled in only a few countries.

Avoiding by-products and emissions

The product obtained from a range of chemical processes has several isomeric forms, i.e. the compounds have the same atomic composition but a different spatial structure. Often only one of the isomers is the 'useful product' and the others are discarded as waste.

For example, *para*-phenylene-diamine (PPD) is produced in considerable quantities as an intermediate for high-performance aramid fibres, along with an equal volume of its isomer, *ortho*-phenylene-diamine (OPD). In the 1980s, aramid fibre producer AKZO-Nobel had no use for OPD. But, as the company needed large quantities of PPD, the development of a selective process to produce only this isomer became a key research target.

Refining creates its own waste. Traditional oil refineries had little scope for controlling the relative fractions of petrol, heavy oils and tar they produced. As there was little or no demand for the heavy fractions (by-products), which contained high concentrations of sulphur, these had to be sold at very low prices. There was also a strong economic driver to control the product fractions emerging from the refinery process. For the refineries, catalytic conversion of tar-like and heavy oil fractions was a means of increasing profits. It also meant they could tackle the high sulphur content of these fractions during conversion. Improved process control technologies permitted more efficient conversion of scarce raw materials into useful end-products and less pollution.

However, economic incentives are not always sufficient. Until the early 1970s, the polyvinyl chloride (PVC) industry emitted considerable quantities of vinyl chloride (VC) feedstock. It was then discovered that vinyl chloride was a carcinogen (affecting mainly the workers within the plants).[8] Under government pressure, processes had to be tightly sealed and maintenance schemes adapted to bring emissions down to acceptable levels (Figure 10.3).

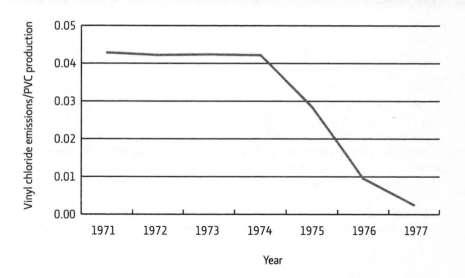

FIGURE 10.3 Reduction of vinyl chloride emissions from the German PVC industry, 1971–1977

8 See the documentary produced by Paola Bonaldi, *Un hombre contra una Industria*, regarding labour actions against the Montecatini/Montedison/EVC Porto Marghera PVC plant near Venice.

Many production processes use a range of solvents, cleaning agents, filtration media and other auxiliaries that do not appear in the end-product. Although these substances are sometimes recycled or re-used, they end up as waste or emissions. As these substances are not part of the core production process, their use is often not given much thought.

> The microelectronics industry had never considered an alternative for CFC113, which was used for cleaning printed circuit boards. Despite its protests, however, the industry was forced by the 1987 Montreal Protocol to seek alternatives. The effect was astonishing: the industry discovered that 70% of its CFC113 use was in fact entirely superfluous and the rest could be avoided almost completely by cleaning the circuit boards with soap and water and then drying them. This solution was in fact rather counter-intuitive, as electronics engineers are generally trained not to use water, because it corrodes connectors. Ford estimated cost reductions of US$ 16 million annually due to the banning of CFC113 in their production processes.[9]

The backcasting method

Optimisation of existing technologies is important but the potential for improvement is often limited. For example, efficiency of electricity production by the burning of fossil fuels (nowadays 40–60%) cannot be improved more than twofold as the energy content of the fuels sets a limit on improvement. If we move to other technological systems, the potential for improvement might be much larger. However, we have to address the needs for which products or services are developed in order to be able to achieve improvements.

Backcasting is a methodological approach for sustainable innovations that starts with analysing needs. It aims to generate long-term options for innovations and to obtain stakeholder consensus regarding those options.

Backcasting (Figure 10.4) consists of:

● Analysing needs

● Identifying options for improvement

● Creating a common future vision with stakeholders

9   K. Mulder, 'Innovation by Disaster: The Ozone Catastrophe as Experiment of Forced Innovation', *International Journal of Environment and Sustainable Development* 4.1 (2005): 88-103.

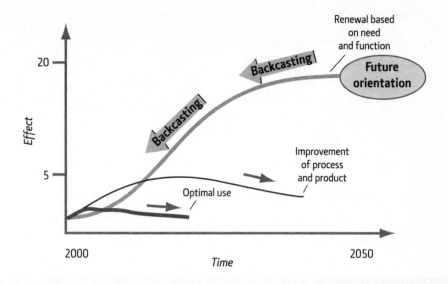

FIGURE 10.4 **Backcasting**

- Developing pathways that could lead to this vision
- Developing consensus on these pathways

Analysing needs and identifying options

An important first step in backcasting is the analysis of need. Before searching for sustainable options to provide for a specific need, it is advantageous to start analysing at the basic level of the need.

Needs may change over time but it is unlikely that basic needs will disappear. Basic needs include:

- Food
- Shelter
- Clean air
- Safety
- Clothing
- Water
- Human communication
- Health
- Self-esteem

- Transport

To provide for basic needs, we generally use at least:

- Energy
- Materials
- Space
- Education

Therefore, these four elements are added to the list of basic needs.

Identifying options for improvement

The next step is to organise a brainstorming session to generate options to fulfil the need. The brainstorm should focus on needs in order to prevent mere extrapolation of current technology.

Options are needed for leapfrogging. Common principles of leapfrogging technologies include:

- Optimise a system first. Only afterwards optimise specific elements of the system such as products and processes

- Minimise waste. It not only saves resources, but waste prevention also saves labour and management time

- Close loops, i.e. try to develop technologies in such ways that products at the end of their life can easily be taken apart and re-used/recycled

- Organise production and consumption in such a way that renewable energy can be used most efficiently

- Use as little material as possible in your design. The less material, the less need there is for resource consumption

- Minimise any damage to ecosystems. Prevent the use of resources that are already consumed in excess of the recovery capacity of planet Earth

- Introduce flexibility in the technological options. Unforeseen events will undoubtedly occur, which will create a need to adapt the innovations that you are pursuing

Creating a common future vision

A common future vision of stakeholders could be a very powerful element in starting to work towards more sustainable options. Powerful future visions were present in the books of Jules Verne. Science fiction such as *Star Trek* also contains a powerful future vision. Nanotechnology is often shown as small machines entering your body to detect and attack any virus. Such future visions are often used to legitimise research. They also guide the choices that researchers make.

Could we reach a common future vision from the various options for improvement? Analysing from a long-term perspective has a major advantage: if stakeholders are challenged to reason from a long-term perspective, they are less likely to focus on their own direct interests. In discussing long-term goals, actors are less reluctant to acknowledge the legitimacy of each other's interests. Moreover, longer-term interest are less fixed and not yet strictly defined. Thus, future visions or guiding visions may result from this process.

Developing pathways towards the common future vision

Consensus on a future vision does not always imply that the stakeholders jointly start to work in the same direction. Actors may very well acknowledge that they should be working in a sustainable direction but that current circumstances are driving them in a different direction. A **socio-technical map** can play a role in identifying promising pathways towards that joint vision. Such a map can help us to identify joint interests and possibilities for compromise.

The socio-technical map

This encompasses:

- The state of development of a technology
- The dynamics in development of this technology
- The different stakeholders involved in this technology
- The views and interest that the stakeholders have in regard to this technology

Developing these pathways resembles **scenario analysis**—a common tool in industrial strategic decision-making. In a scenario, a possible future is portrayed. The scenario should be credible and tantalising as a possible development, and should therefore be consistent and sufficiently detailed.

For scenario analysis, we need at least three scenarios. The most important goal of scenarios is not to predict but to 'wake people up' and make them aware of possible changes. Scenarios often comprise the input for a creative workshop.

Backcasting differs from scenario analysis in that ordinary scenarios are coherent forecasts of several paths into the future, while backcasting is trying to identify paths that finish at a specific end-situation.

A scenario attempts to plot the choices or key events and to translate the consequences of a choice or event into later choices or events (a choice often involves the elimination of a later possibility).

> During stable times, the mental model of a successful decision-maker and unfolding reality match . . . In times of rapid change and increased complexity, however, the manager's mental model becomes a dangerously mixed bag: rich detail and understanding can co-exist with dubious assumptions and illusory projections.[10]

Trend scenarios show developments that are in line with our current ideas. They are also called 'surprise-free scenarios' because they do not incorporate any sudden and unexpected events. The scenarios are normally shown as surrounding a most probable scenario (which often represents 'business as usual').[11]

Backcasting is in this respect the reverse of scenarios: it analyses which paths may be able to lead us to the future vision and tries to derive crucial decisions that we have to take to be able to reach (one of) the right track(s).

Examples

Plastics

What are the possible options for a sustainable future for plastic materials? Plastics are a class of products that can be said to be unsustainable in several respects:

- They deplete non-renewable resources such as crude oil (formed over millions of years)
- They generate a variety of emissions
- They create a waste stream of plastic end-products

One could say that the main problem with plastics is that we are rapidly transforming crude oil into litter, non-degradable waste and carbon dioxide (CO_2). However, plastics are lightweight materials and therefore efficient in use. They also do not corrode. We could speak of a sustainable plastics 'cycle' if we could:

10 P. Wack, 'Scenarios: Uncharted Waters Ahead', Harvard Business Review, September/October 1985: 73-89.
11 P. Rademaker, 'Toekomstverkenning in het bedrijfsleven', in J. van Doorn and F. van Vught (eds.), Nederland op zoek naar zijn toekomst [The Netherlands Searching for its Future] (Utrecht: Spectrum, 1981): 170-89.

- Produce plastics solely from biomass
- Re-use plastics materials and products
- Burn the remaining waste to recover its energy content

Figure 10.5 indicates the current plastics cycle in which oil is transformed into the plastic product. This plastic product fulfils our needs. The main problems derive from the fact that plastic is transformed only very slowly into its initial state, i.e. crude oil.[12]

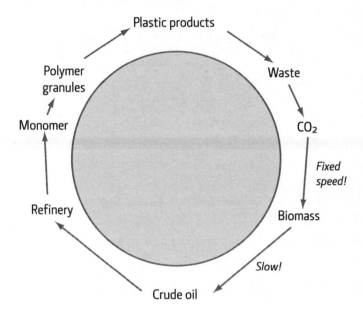

FIGURE 10.5 Current cycle for plastics

For sustainable development, we want to fulfil the need with the product but we also want to close the loops around the product as much as possible in order to be most efficient. Durable products and product re-use are thus preferred. Recycling plastic materials and transforming plastic waste into chemical feedstock contribute to the solution. Using biomass as the raw material creates a shortcut across the slowest part of the cycle (Figure 10.6).

Several initiatives have been taken to develop these solutions. Recycling of plastics to feedstock has been introduced in some countries and biomass routes to produce raw materials are being studied. These options diminish

12 The problem would remain the same if we could make plastics based on coal or natural gas.

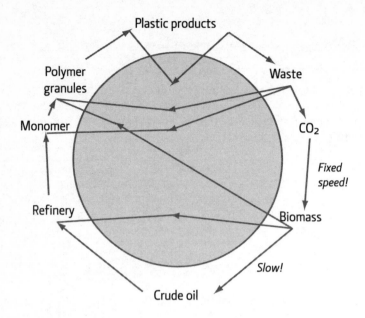

FIGURE 10.6 **A sustainable future for plastics?**

raw materials consumption, which does not have a major economic impact on existing companies. However, if we target a reduction in the consumption of plastics, various interests could be affected.

Energy

Our energy supply systems will be key elements on the road to sustainable development. Many other production systems will move closer towards sustainability if:

- We are able to supply the world using renewable primary energy sources
- Do not endanger local or global ecosystems (during normal operation as well as through abnormal operation and accidents)
- Enable local communities to exert control over crucial elements of their societies

The main renewable forms of energy are:

- Solar (see Figure 10.7)
- Wind

FIGURE 10.7 **Solar energy: solar panel**

- Biomass
- Hydropower
- Wave
- Ocean heat[13]
- Geothermal
- Tidal

Some people would like to add nuclear fission and nuclear fusion to this list of sustainable energy sources. However, it is rather doubtful whether nuclear fusion, using the vast amounts of deuterium available in the water on the planet, can ever become more than a scientific curiosity; after half a century of research, the odds are against it. Nuclear fission has left us with

13 Using the differences in temperature with upper layers of the atmosphere.

large amounts of radioactive waste and the Chernobyl catastrophe that will have a death toll for centuries to come.[14]

Items 2-6 on the list of renewable energies are actually all forms of solar energy. The Sun supplies the Earth with about 1,500 times the power used by society today. There is therefore ample renewable energy available; the main problem is how to use it effectively. Further optimisation of technologies is needed to decrease costs and boost efficiency. Under favourable conditions, however, it is estimated that technologies could be developed that would generate electricity from these sources at a cost of US$0.02–0.06 per kWh.

Every form of energy conversion is associated with losses and costs. For example, solar or geothermal heat can be used most effectively for heating, and wind for direct propulsion of shipping vessels. Given the dependence on natural circumstance, renewable sources of energy should be combined prudently in systems in order to use them effectively.

Storage

Solar, wind, tidal and wave energy cannot be tapped when and as we chose. Storage is therefore required. If solar energy is collected as heat, it can be stored as hot water. If electricity is to be generated, however, storage becomes a problem. This can be solved:

- By converting the electricity into other forms of energy; for instance, by:
  - Pumping water to high reservoirs and using it later as hydropower
  - Making hydrogen, which can be used in fuel cells to generate electricity
  - Charging batteries
  - Increasing pressure behind pressure valves

- By integrating these sources of energy into a more efficient electricity supply system. An individual country may sometimes be cloudy but never an entire continent. Likewise, strong winds in one country can compensate for calm weather elsewhere.[15] Hydropower

14 The International Commission on Radiation Protection (ICRP) estimates that there are 0.04 cancer deaths for each person-Sievert. As it is estimated that the Chernobyl accident created an exposure in Europe and the former USSR of 600,000 person-Sievert, there will be about 24,000 additional cancer deaths. See: R. Garwin, 'More on Deaths due to Chernobyl', *Physics and Society* 28.1 (1999); www.fas.org/rlg/ljan99.html, 14 November 2005.

15 P. Dowling, 'Europe Needs a Super Grid', paper presented at the European Wind Energy Conference held in London, November 2004; www.ewea.org/documents/WD24ii_FOCUS.pdf, 14 November 2005.

could compensate for temporary shortages of these non-controllable energy sources

However, all the forms of storage suggested above are expensive and are associated with considerable losses as double conversion is required.

Ecology

Hydropower often ruins local ecology. Valleys are turned into lakes and barrier dams also act as barriers to wildlife such as salmon. Millions of people may have to be displaced as is the case with the Three Gorges Dam in China.[16] Over time, the retaining lakes may become less effective as they fill up with plants and sediment. They may also become a breeding ground for insects.

Wind turbines are often criticised for killing birds. However, if sites are chosen carefully to avoid seasonal migration routes, the number of bird casualties can be kept very low. Windmill noise and shade may be serious problems, but only in the direct vicinity. Prudent site selection is therefore essential.

Risks

In the event of an earthquake or military conflict, the lakes required for hydropower represent an enormous risk; if a dam collapsed, the enormous rush of water could kill millions of people.

The same may hold for notions of utilising ocean heat. This could probably only be achieved by erecting enormous towers evaporating low-boiling point liquids such as ammonia and generating electricity by collecting the condensed liquid at the top of the tower. In the event of an earthquake or an aircraft collision, an enormous disaster could ensue.

Space

Because it is readily stored, biomass is the preferred energy source for many people. However, the traditional burning of biomass carries considerable health risks as it often causes severe indoor air pollution.[17]

If the world's population increases to 9–10 billion, more agricultural land will be needed to produce sufficient food and materials. As the most productive land is already used for farming, a 50% rise in population could necessitate a far greater increase in agricultural land use unless agricultural efficiency is improved substantially. Could more agricultural land be used for energy generation? This would undoubtedly imply further damage to ecosystems.

16 Dai Qing, *The River Dragon Has Come! The Three Gorges Dam and the Fate of China's Yangtze River and its People* (Armonk, NY: M.E. Sharpe Inc., 1998).
17 www.epa.gov/iaq/pcia.html, 17 March 2006.

There are many promising technologies with which to create sustainable energy supply systems. The main challenge will be to create a reliable energy system based on renewable sources that can effectively overcome these drawbacks, while enabling local and regional communities to regain or maintain control of their energy systems.

Alternatives to meat

The production and consumption of meat has huge environmental impacts. Large areas of tropical forests are cleared to produce soy fodder to feed the livestock of farmers in Europe and North America. Livestock breeding is a concentrated activity and it causes environmental destruction in its main areas. Locally it creates acidification, destruction of ecosystems and eutrophication[18] and it contributes to climate change. Soy fodder, the animals and the meat are often transported over large distances.

The main need that is fulfilled by eating meat is the need for protein. However, animals are not very efficient in converting fodder crops into proteins. About 6 kg of plant proteins are needed to produce 1 kg of animal proteins. To produce 1 kg of beef, 15 m^3 of water is needed. For 1 kg of lamb, 10 m^3 is needed. However, it takes only 0.4–3 m^3 of water to produce 1 kg of wheat.

Meat consumption is growing rapidly. While the world population grew from 2.7 billion to 6 billion in the past 50 years, meat production increased fivefold. However, meat consumption also fulfils other functions such as tradition or expression of status. Meat production is an important economic activity that contributes to the livelihood of the countryside.

Protein food products could also be produced using vegetable proteins. In principle, this makes protein production about six times more efficient. Various protein products have been developed such as different varieties of tofu and textured vegetable protein (TVP). However, these vegetarian foods have a rather negative image for mainstream consumers due to the poor quality and the association with alternative lifestyles and radical vegetarianism.

Novel protein foods (NPFs) based on plant proteins are being developed as a superior alternative. Various efforts have been made to develop protein foods that are attractive to both consumers and producers, and which can fulfil the same dietary and cultural role as meat.[19]

Scandals in the meat supply chain (BSE,[20] foot and mouth disease, bird flu and the massive slaughtering to control these diseases, etc.) have furthered the impetus of NPFs. Plant protein products are being introduced onto the

18 High nutrient concentrations in aquatic ecosystems stimulate blooms of algae, thereby ruining the ecosystem.
19 Protein Foods, Environment, Technology and Society (PROFETAS); www.profetas.nl, 14 November 2005.
20 Bovine spongiform encephalopathy (also known as mad cow disease).

market, but it will not be easy to reverse the trend of rapidly growing meat consumption—especially in countries where affluence is increasing.

Closing remarks

Technologies can and will have a major role to play in sustainable development, but never as the 'Deus ex machina' that will relieve us of our problems at no cost and without requiring any further adaptation. Solutions will always be 'socio-technical' in nature; in other words, they will encompass technological as well as social transformation.

Two words of warnings are in order here:

- Improved technology may be the enemy of truly sustainable technology. Studies of technological change teach us that the development of technology is a path-dependent process, i.e. we cannot choose freely but are bound by the choices of the past. Similarly, our choices today have consequences for the range of options left open to future generations. As a consequence, what today may appear to be an improvement may in fact mean embarking on a technological strategy that prevents us choosing an even better solution later on—a phenomenon known as 'lock-in'

- Striving only for truly sustainable technologies may stop us taking action now. Solving the problems we face today will hurt. Doubts about the viability of particular solutions often serve merely as an excuse for postponing action. In practice, the precautionary principle is frequently reversed: as long as the existence and gravity of the problem has not been proven beyond doubt, action is deferred. This is an attitude we can no longer afford to adopt. To develop new technologies takes time. Even development of a new car, involving no revolutionary new technology, takes about ten years. Revolutionary changes will take decades of concerted action on the part of governments, consumers and industry—particularly the vehicle and energy supply industries

It is important for society to try to foresee the impact of technological change, so that the merits of such change can be discussed democratically. However, experience with parliamentary technology assessment during the last few decades of the 20th century has shown that we can gain no more than a limited insight into the future impact of technologies because that impact is intimately bound up with—or even indistinguishable from—wider, more general cultural changes. Unforeseen rebound effects and new social dilemmas occur frequently. They cannot be prevented and it is therefore crucial that technological strategies towards sustainability are flexible.

Our challenge today is to learn from past mistakes and correct what we can with due haste. And to start doing so now.

Questions, discussion and exercises

Hybrid cars

Toyota sells the hybrid car Prius. Two engines propel this car; one is electric for driving in metropolitan areas and the other is a diesel engine for higher speeds on motorways. The batteries that power the electric engine are recharged during diesel propulsion. The overall fuel consumption is 1 litre of diesel for every 19 kilometres of travel.

a. For which environmental problem is this car the solution?

b. How could hybrid cars contribute to making a leap in the environmental efficiency of car transport?

Future of metals

Develop a future vision to provide steel or aluminium in a sustainable way. Which new technologies should be developed? What is the main difference from a sustainable plastics cycle?

Future of textiles

Develop a future vision to provide textiles in a sustainable manner. Which new technologies should be developed? What is the main difference from a sustainable plastics cycle?

Meat and culture

Meat consumption is strongly related to culture. Several religions rule out the consumption of specific types of meat or prescribe meat consumption at specific events. Find out in which religions NPFs have the best chance of being accepted. In which religions might objections to NPFs be stronger?

Backcasting (1)

Iceland is planning to convert its economy entirely to renewable energy in 25 years. It has plenty of geothermal energy and hydropower as primary energy sources. Which intermediate steps should Iceland take to achieve this goal? What is the importance of this development in Iceland for other nations?

Backcasting (2)

Suppose that there is consensus on the vision that your country should be able to be energy-self-sufficient in 50 years without using any fossil fuels. Derive a number of measures that should be taken and identify some milestones.

Backcasting (3)

Three visions regarding future food consumption are shown in the box below. Discuss them. Start backcasting from these future visions and identify a number of measures to be taken. Give indications regarding the timing of measures and address the issue of stakeholder involvement.

Future food consumption

As Susan and Erik hardly ever eat their dinner at home, they have little food waste. But the Super-Rant (Figure 10.8) produces lots of waste. The food waste and biodegradable packaging from the houses and the Super-Rant are collected and used for regional energy production.

Dishwashing at the Super-Rant is done on a larger scale with professional equipment. All plates that are taken home by Susan and others are also cleaned at the Super-Rant *(continued over)*.

FIGURE 10.8 **The 'Super-Rant'**

Drawing by Peter Welleman

The Casper family (Figure 10.9) has a dishwasher. Packaging waste is biodegradable and it is collected together with food waste before it is used for regional energy production. The chips in the packaging can be recycled. Water and energy used by the Casper family for cooking and food storage is re-used in the house as much as possible.

Hedi and Jan (Figure 10.10) do not have a dishwasher and do the dishwashing by hand as they think it is only a minor task. Food waste and biodegradable packaging are composted in the garden and the compost is used in their garden. Non-biodegradable packaging mostly offers a deposit, so they take it back to the shop from which they bought it.

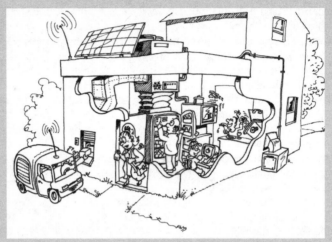

FIGURE 10.9 **The Casper family's house**

Drawing by Peter Welleman

FIGURE 10.10 **Hedi and Jan in their garden**

Drawing by Peter Welleman

Suggestions for further reading

Allen, D.T., and K. Sinclair, *Pollution Prevention for Chemical Processes* (New York: John Wiley, 1997).

Ausubel, J.H., and H.E. Sladovich, *Technology and Environment* (Washington, DC: National Academy Engineering, National Academy Press, 1989).

Brauni, E., *Futile Progress: Technology's Empty Promise* (London: Earthscan, 1995).

Corbitt, R.A. *Standard Handbook of Environmental Engineering* (New York: McGraw-Hill, 1999).

Graedel, T.E. and B.R. Allenby, *Industrial Ecology* (Englewood Cliffs, NJ: Prentice Hall, 1995).

Hawken, P., and L.H. Lovins, *Natural Capitalism: Creating the Next Industrial Revolution* (Snowmass, CO: Rocky Mountains Institute, 1999).

Kemp, R., 'Technology and the Transition to Environmental Sustainability: The Problem of Technological Regime Shifts', *Futures* 26.10 (1994): 1,023-46.

Kirkwood, R.C., and A.J. Longley, *Clean Technology and the Environment* (London: Blackie Academic & Professional, 1995).

Leal, F.W., *Environmental Engineering: International Perspectives* (Frankfurt/Berlin/New York/Paris/Vienna: Peter Lang, 1998).

Moors, E., *Metal Making in Motion: Technology Choices for Sustainable Metals Production* (Delft, the Netherlands: Delft University Press, 2000).

Moors, E., and K.F. Mulder, 'Industry in Sustainable Development: The Contribution of Regime Changes to Radical Technical Innovation in Industry', *International Journal of Technology Policy and Management* 2.4 (2002): 434-54.

Mulder, K., 'Sustainable Consumption and Production of Plastics?', *Technological Forecasting and Social Change* 58.1/2 (1998): 105-24.

Nazaroff, W.W., and L. Alvarez Cohen, *Environmental Engineering Science* (New York: John Wiley, 2001).

Schipper, L., and M. Grubb, 'On the Rebound: Between Energy Intensities and Energy Use in IAEA Countries', *Energy Policy* 28 (2000): 367-88.

Schmidt-Bleek, F., *MIPS and Factor 10 for a Sustainable and Profitable Economy* (Wuppertal, Germany: Wuppertal Institute, 1997).

Tierny, R.J., 'Green by Design: Factor 10 goals at Pratt and Whitney', *Corporate Environmental Strategy* 9 (2002): 52-61.

Turkenburg, W.C., 'Renewable Energy Technologies' in the *World Energy Assessment* (New York: United Nations Development Programme, 2000): ch. 7.

Von Weizsäcker, E.U., A. Lovins and L.H. Lovins, *Factor Four: Doubling Wealth—Halving Resource Use* (London: Earthscan, 1997).

Weaver, P., L. Jansen, G. van Grootveld, E. van Spiegel and P. Vergragt, *Sustainable Technology Development* (Sheffield, UK: Greenleaf Publishing, 2000).

Index